2/18/16
$ 105.00

## Science, Technology and Medicine in Modern History

General Editor: John V. Pickstone, Centre for the History of Science, Technology and Medicine, University of Manchester, England (www.man.ac.uk/CHSTM)

One purpose of historical writing is to illuminate the present. At the start of the third millennium, science, technology and medicine are enormously important, yet their development is little studied.

The reasons for this failure are as obvious as they are regrettable. Education in many countries, not least in Britain, draws deep divisions between the sciences and the humanities. Men and women who have been trained in science have too often been trained away from history, or from any sustained reflection on how societies work. Those educated in historical or social studies have usually learned so little of science that they remain thereafter suspicious, overawed, or both.

Such a diagnosis is by no means novel, nor is it particularly original to suggest that good historical studies of science may be peculiarly important for understanding our present. Indeed this series could be seen as extending research undertaken over the last half-century. But much of that work has treated science, technology and medicine separately; this series aims to draw them together, partly because the three activities have become ever more intertwined. This breadth of focus and the stress on the relationships of knowledge and practice are particularly appropriate in a series which will concentrate on modern history and on industrial societies. Furthermore, while much of the existing historical scholarship is on American topics, this series aims to be international, encouraging studies on European material. The intention is to present science, technology and medicine as aspects of modern culture, analysing their economic, social and political aspects, but not neglecting the expert content which tends to distance them from other aspects of history. The books will investigate the uses and consequences of technical knowledge, and how it was shaped within particular economic, social and political structures.

Such analyses should contribute to discussions of present dilemmas and to assessments of policy. 'Science' no longer appears to us as a triumphant agent of Enlightenment, breaking the shackles of tradition, enabling command over nature. But neither is it to be seen as merely oppressive and dangerous. Judgement requires information and careful analysis, just as intelligent policy-making requires a community of discourse between men and women trained in technical specialities and those who are not.

This series is intended to supply analysis and to stimulate debate. Opinions will vary between authors; we claim only that the books are based on searching historical study of topics which are important, not least because they cut across conventional academic boundaries. They should appeal not just to historians, nor just to scientists, engineers and doctors, but to all who share the view that science, technology and medicine are far too important to be left out of history.

### Titles include:

Julie Anderson, Francis Neary and John V. Pickstone
SURGEONS, MANUFACTURERS AND PATIENTS
A Transatlantic History of Total Hip Replacement

Roberta E. Bivins
ACUPUNCTURE, EXPERTISE AND CROSS-CULTURAL MEDICINE

Linda Bryder
WOMEN'S BODIES AND MEDICAL SCIENCE
An Inquiry into Cervical Cancer

Roger Cooter
SURGERY AND SOCIETY IN PEACE AND WAR
Orthopaedics and the Organization of Modern Medicine, 1880–1948

Jean-Paul Gaudillière and Ilana Löwy (editors)
THE INVISIBLE INDUSTRIALIST
Manufacture and the Construction of Scientific Knowledge

Jean-Paul Gaudillière and Volker Hess (editors)
WAYS OF REGULATING DRUGS IN THE 19TH AND 20TH CENTURIES

Christoph Gradmann and Jonathan Simon (editors)
EVALUATING AND STANDARDIZING THERAPEUTIC AGENTS, 1890–1950

Sarah G. Mars
THE POLITICS OF ADDICTION
Medical Conflict and Drug Dependence in England since the 1960s

Alex Mold and Virginia Berridge
VOLUNTARY ACTION AND ILLEGAL DRUGS
Health and Society in Britain since the 1960s

Ayesha Nathoo
HEARTS EXPOSED
Transplants and the Media in 1960s Britain

Neil Pemberton and Michael Worboys
MAD DOGS AND ENGLISHMEN
Rabies in Britain, 1830–2000

Cay-Rüdiger Prüll, Andreas-Holger Maehle and Robert Francis Halliwell
A SHORT HISTORY OF THE DRUG RECEPTOR CONCEPT

Thomas Schlich
SURGERY, SCIENCE AND INDUSTRY
A Revolution in Fracture Care, 1950s–1990s

Eve Seguin (editor)
INFECTIOUS PROCESSES
Knowledge, Discourse and the Politics of Prions

Crosbie Smith and Jon Agar (editors)
MAKING SPACE FOR SCIENCE
Territorial Themes in the Shaping of Knowledge

Stephanie J. Snow
OPERATIONS WITHOUT PAIN
The Practice and Science of Anaesthesia in Victorian Britain

Carsten Timmermann and Julie Anderson (editors)
DEVICES AND DESIGNS
Medical Technologies in Historical Perspective

Carsten Timmermann and Elizabeth Toon (editors)
CANCER PATIENTS, CANCER PATHWAYS
Historical and Sociological Perspectives

Duncan Wilson
TISSUE CULTURE IN SCIENCE AND SOCIETY
The Public Life of a Biological Technique in Twentieth Century Britain

---

**Science, Technology and Medicine in Modern History**
**Series Standing Order ISBN 978–0–333–71492–8 (hardcover)**
**Series Standing Order ISBN 978–0–333–80340–0 (paperback)**
*(outside North America only)*

You can receive future titles in this series as they are published by placing a standing order. Please contact your bookseller or, in case of difficulty, write to us at the address below with your name and address, the title of the series and one of the ISBNs quoted above.

Customer Services Department, Macmillan Distribution Ltd, Houndmills, Basingstoke, Hampshire RG21 6XS, England

---

*Also by Carsten Timmermann*

DEVICES AND DESIGNS: Medical Technologies in Historical Perspective (*edited with Julie Anderson*)
THE RECALCITRANT DISEASE: A History of Lung Cancer (forthcoming)

*Also by Elizabeth Toon*

PRIVATE TRAUMA, PUBLIC DRAMA: Breast Cancer Treatment in Twentieth-Century Britain (forthcoming)

# Cancer Patients, Cancer Pathways

## Historical and Sociological Perspectives

Edited by

Carsten Timmermann and Elizabeth Toon
*University of Manchester*

First published 2012 by
PALGRAVE MACMILLAN

Palgrave Macmillan in the UK is an imprint of Macmillan Publishers Limited, registered in England, company number 785998, of Houndmills, Basingstoke, Hampshire RG21 6XS.

Palgrave Macmillan in the US is a division of St Martin's Press LLC, 175 Fifth Avenue, New York, NY 10010.

Palgrave Macmillan is the global academic imprint of the above companies and has companies and representatives throughout the world.

Palgrave® and Macmillan® are registered trademarks in the United States, the United Kingdom, Europe and other countries

ISBN 978-1-137-27207-2

This book is printed on paper suitable for recycling and made from fully managed and sustained forest sources. Logging, pulping and manufacturing processes are expected to conform to the environmental regulations of the country of origin.

A catalogue record for this book is available from the British Library.

A catalog record for this book is available from the Library of Congress.

# Contents

# List of Figures

# Acknowledgements

The editors would like to thank the Wellcome Trust for their support of the Constructing Cancers project (Programme Grant 068397), which funded the research of several contributors to this volume, as well as the conference where most of these essays were originally presented. Our thanks also go to the scholars who have joined us in an informal 'cancer history' network, for their many stimulating discussions in Manchester, Paris, Bethesda and elsewhere. We greatly appreciate the critiques, input and support given by our colleagues at Manchester's Centre for the History of Science, Technology and Medicine; thanks especially are due to Annie Jamieson for her editorial assistance with the final manuscript. Our greatest thanks, though, go to the contributors to this volume, for their patience and good cheer throughout its very long gestation period.

The editors and publishers wish to thank the following for permission to reproduce copyright material:

- Oxford University Press, for 'Schematic Depiction of a "Normal" Treatment Pathway for Non-Small Cell Lung Cancer', Chris Williams, *Lung Cancer: The Facts*, 2nd edition (1992), p. 51.
- The Royal Marsden NHS Foundation Trust, for the first panel of the original comic, *Captain Chemo*, 1999, and the website version of *Captain Chemo*, 2001.
- The Leukaemia Research Fund, for the depiction of Mr. Wiggly from *Jack's Diary*.
- The University of Chicago Press, for material in Gerald Kutcher's chapter, previously published by Kutcher in *Contested Medicine: Cancer Research and the Military* (2009).
- Terry Dennett and the Jo Spence Memorial Archive, for Jo Spence, *Patient's Eye View* (1989).

# Notes on the Contributors

**Joanna Baines** completed her doctorate on the individualisation of cancer at the Centre for the History of Science, Technology and Medicine, University of Manchester in 2010. She then took a public engagement role and is currently co-authoring a book on the history of childhood cancer in Britain.
Email: joanna.baines-2@manchester.ac.uk

**Roland Bal** is Professor and Chair of the Department of Healthcare Governance of the Institute of Health Policy and Management in Rotterdam, the Netherlands. His research interests include quality and safety of care and governance infrastructures in health care. He has published extensively on the development and application of information technologies and on the creation of public accountabilities in health care and is leading the evaluation of several large-scale quality of care programmes in the Netherlands. He was a member of the board of the 'Better Faster' quality collaborative in the Netherlands, a programme in which 24 Dutch hospitals participated to create innovative health care organisations, and leads the Dutch part of the EU funded Quaser project on quality management in European hospitals.
Email: r.bal@bmg.eur.nl

**Marc Berg** is a partner at the health care consultancy agency Plexus Medical Group and affiliated with the Institute of Health Policy and Management, Erasmus University Medical Center. His main interests are the generation and organisation of 'high quality' and 'low cost' health care practices, both at the level of the health care practices itself, as well as on the system level. Marc has co-founded the largest Dutch hospital redesign project in the Netherlands, *Better Faster*, which has been implemented in 25 per cent of Dutch hospitals. He is the Chairman of the Dutch Society for Quality in Health Care, and is deeply involved in policy debates on the future of health care systems. He helps insurance companies creating novel and constructive strategies in the emerging health care market, and works as coordinator for the Dutch Health Care Research Fund (ZonMw). Marc has published widely on medical sociology, sociology of technology, standardisation, information technology and quality management. His books include *Rationalizing Medical Work: Decision Support Techniques* *and Medical Practices* (1997) and (with Stefan Timmermans) *The Gold Standard: The Challenge of Evidence-Based Medicine and Standardization in Health Care* (2003).

**Alberto Cambrosio** is a Professor in the Department of Social Studies of Medicine at McGill University. His present work focuses on the intersection between genomics and cancer diagnosis and therapy. He has co-authored (with Peter Keating) *Exquisite Specificity: The Monoclonal Antibody Revolution* (1995), *Biomedical Platforms: Realigning the Normal and the Pathological in Late-Twentieth-Century Medicine* (2003), and *Cancer on Trial: Oncology as a New Style of Practice* (2012).
Email: alberto.cambrosio@mcgill.ca

**Emm Barnes Johnstone** works at Royal Holloway, University of London, combining public engagement programming and history of medicine research and teaching. She has written on the history of childhood cancer and patient support services, including 'Cancer Coverage: The Public Face of Childhood Leukaemia in 1960s Britain' (*Endeavour*, 2008) and *The Changing Faces of Childhood Cancer* (forthcoming), co-authored with Joanna Baines. She has also published *The Art of Medicine: Over 2,000 Years of Images and Imagination* (2011), co-authored with Julie Anderson and Emma Shackleton.
Email: Emm.Johnstone@rhul.ac.uk

**Peter Keating** is Professor of History at the Université du Québec à Montréal where he is a member of the Centre interuniversitaire de recherche sur la science et la technologie. He has co-authored (with Alberto Cambrosio) *Exquisite Specificity: The Monoclonal Antibody Revolution* (1995), *Biomedical Platforms: Realigning the Normal and the Pathological in Late-Twentieth-Century Medicine* (2003) and *Cancer on Trial: Oncology as a New Style of Practice* (2012). He is presently working on the development of genomic medicine.
Email: keating.peter@uqam.ca

**Gerald Kutcher** is Professor in the Department of History at Binghamton University in New York and an Affiliated Research Scholar in the Department of History and Philosophy of Science at the University of Cambridge. Problematic medical research was the subject of his first book, *Contested Medicine: Cancer Research and the Military* (2009). He is currently a Simon Guggenheim Fellow working on his next book with the working title *High Expectations: A History of Cancer Therapies*.
Email: gkutcher@binghamton.edu

**Ilana Löwy** is a historian of medicine and a senior researcher at INSERM (Institut national de la santé et de la recherche médicale, France). Her research focuses on interactions between laboratory sciences, clinical medicine and public health. She studied the history of the 'Pasteurian sciences', tropical medicine, oncology, and women's reproductive health, and is now investigating intersections between the history of contraception, the prevention of STDs, and the history of prenatal diagnosis. Her books include: *Between Bench and Bedside: Science, Healing and Interleukin 2 in a Cancer Ward* (1996), *Preventive Strikes: Women, Precancer and Prophylactic Surgery* (2009), and *A Woman's Disease: The History of Cervical Cancer* (2011). Email: lowy@vjf.cnrs.fr

**John Pickstone** is Emeritus Professor of the History of Knowledges at the University of Manchester. He has worked in Manchester since 1974, and in 1986 he set up the Wellcome Unit and the Centre for the History of Science, Technology and Medicine. He is the author of *Ways of Knowing: A New History of Science, Technology and Medicine* (2001); editor with Roger Cooter of the *Companion to Medicine in the Twentieth Century* (2002); author with Julie Anderson and Francis Neary of *Surgeons, Manufacturers and Patients: A Transatlantic History of Total Hip Replacement* (2007); and editor with Peter Bowler of Volume 6 of the *Cambridge History of Science* (2009). He continues to do research on cancer history, the recent NHS, Manchester science, technology and medicine, and 'working knowledges' (*Isis*, 2007).
Email: john.pickstone@manchester.ac.uk

**Carsten Timmermann** is a Lecturer at the Centre for the History of Science, Technology and Medicine, University of Manchester. He was lead researcher of the Wellcome Trust-funded research programme 'Constructing Cancers 1945–2000'. He has published on the debate over a crisis of medicine in Germany in the interwar period, the history of hypertension as an illness defined by statistics, the history of risk factors, and most recently the history of lung cancer. He is the editor (with Julie Anderson) of *Devices and Designs: Medical Technologies in Historical Perspective* (2006) and the author of *The Recalcitrant Disease: A History of Lung Cancer* (forthcoming). Email: carsten.timmermann@manchester.ac.uk

**Elizabeth Toon** is a Temporary Lecturer and former Wellcome Research Associate at the Centre for the History of Science, Technology and Medicine, University of Manchester, and was also a member of the 'Constructing Cancers' research programme at CHSTM. She is currently completing *Private Trauma, Public Drama: Breast Cancer Treatment in Britain, 1920–1985*, under

contract with Palgrave. Her research focuses on expert-lay relationships around biomedical knowledge, and especially on the ways that gender shapes those relationships. As well as her work on breast cancer in twentieth-century Britain, Toon has published on public health programmes and popular health practices in twentieth-century United States. Her latest project, funded by the Wellcome Trust, is a history of technologies and practices around women's cancer screening in the UK.
Email: elizabeth.toon@manchester.ac.uk

**Helen Valier** is Instructional Assistant Professor and Coordinator of the Medicine and Society Program at the Honors College, University of Houston. Until 2005 she was a Research Associate and member of the 'Constructing Cancers' team at the University of Manchester. She is interested in the history of cancer clinical trials, focusing especially on the MD Anderson Cancer Center in Houston, the MRC cancer research programme, and US-UK interactions in cancer research. She is the author (with John Pickstone) of *Community Professions and Business: A History of the Central Manchester Teaching Hospitals and the National Health Service* (2008). Email: hkvalier@uh.edu

**Teun Zuiderent-Jerak** is Associate Professor of Science and Technology Studies at the Department of Healthcare Governance of the Institute of Health Policy and Management at the Erasmus University Rotterdam, the Netherlands. His research focuses on standardisation and clinical practice guideline development in health care, the construction of markets for public goods, and STS research that explicitly aims to 'intervene' in the practices it studies. He has carried out a range of research and organisational change projects. He was project manager for setting up a haemophilia care centre of a large university hospital, project manager of 'Working Differently', a reorganisation project at a large haematology/oncology ward, and project leader of the process redesign project in the *Better Faster* quality collaborative for improving hospital care delivery. During this last project he was responsible for around 40 improvement projects in 16 hospitals which all focused on reducing the throughput time and length of stay for oncology and elective patients. He has published articles in academic journals such as *Social Studies of Science, Science, Technology, and Human Values, Social Science & Medicine,* and *Science as Culture.* His book *Situated Intervention; Sociological Experiments in Healthcare* is forthcoming.
Email: zuiderent@bmg.eur.nl

# 1
# Introduction

*Carsten Timmermann and Elizabeth Toon*

This volume was conceived during a workshop held at the Centre for the History of Science, Technology and Medicine, University of Manchester in October 2005, organised with funding from a Wellcome Trust programme grant, 'Constructing Cancers, 1945–2000'. This was the second of a series of conferences, some of which were hosted by colleagues in the US and in France. Our goal in that workshop, and now in this volume, has been to explore the history of cancer research and treatment as everyday practice, by following patients on their journeys through the institutions of cancer medicine in the West since the Second World War.

Historians of medicine and other writers have discovered cancer as a subject. Recent historical accounts (including work by scholars contributing to this volume) discuss cancer's emergence as a research question, policy problem, and public threat. Some of these accounts chronicle the development of national programmes and specialised institutions, while others analyse popular portrayals of cancer and its treatment in media and culture; still others focus on evolving controversies, both within medicine and in the greater public sphere, around treatment, causation, and prevention.[1] Meanwhile, practitioner accounts have often depicted the history of cancer research, treatment and control as a story of hard-won progress, as they track the accumulation of scientific findings and their application to clinical practice.[2]

But historians and health care professionals have not been the only ones writing about the disease: since the 1980s, a growing number of patients have sought to tell not *the* story of cancer, but their *own individual* cancer stories. Autobiographies, edited diaries, newspaper and magazine columns, films and television shows, comics, and blogs – all are intended by their creators to document their experiences of treatment.[3] For some of these patients, as social scientists have shown, telling one's own story helps them to find meaning in their experience; they embrace the illness as an inherent part of their biographies, and claim new identities as sufferers or survivors.[4] Others facing the disease use the cancer story as a mode of cultural critique; narrating their illness encounters not only allows them to regain control over their experiences, but also to integrate those experiences into a broader political understanding.[5] The image reproduced on this volume's cover and as Figure 1.1 (*Patient's Eye View*, 1989), by socialist feminist photographer Jo Spence in collaboration with David Roberts, does the latter.[6] Spence's photographs, some taken during treatment and others staged afterwards, consciously invoke, as sociologist Susan Bell has written, 'the passive and unknowing position of a woman patient'.[7]

In *Patient's Eye View*, the camera suggests a patient lying on her back, looking up at three scrubbed-up people, one holding a scalpel; their individual identities are masked by professional clothing as they prepare to operate. Spence, who later employed traditional Chinese therapies and embraced alternative therapeutic regimes to manage her cancer, hoped her art would encourage others to question the power relationships inherent within the orthodox medical encounter.[8] Narrative critiques like Spence's encourage us as scholars to ask how we can reorient our work, and try to understand the history of cancer treatment as experienced on both ends of

*Figure 1.1*   Jo Spence, *Patient's Eye View*, 1989. Reprinted with permission.

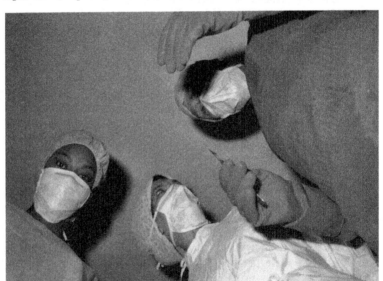

the camera.[9] In the same vein, the accounts of patient experience we present in some of the chapters in this volume are not primarily biographical but rather are intended to reveal the context that has framed and conditioned these experiences.

While patients experience cancer treatment as a unique life-changing series of events, to medical professionals it constitutes routine work practice. Historians of science and medicine, sociologists, and science studies scholars have developed tools that help us better understand how these routines are established and maintained.[10] Rather than telling the story of cancer as that of inevitable progress and obstacles overcome, these approaches emphasise the importance of understanding how contingency, politics, and institutional interests inform approaches to research and treatment. Our collection therefore considers the history of cancer research and treatment as everyday practices embedded in institutional structures that are exemplary of late-twentieth-century medicine. In other words, our focus is on how patient experiences of cancer are shaped through treatment pathways, and how treatment pathways are shaped by the patients who pass through them and the health care professionals who design and execute them.

To examine this interface between the trajectories of individuals and the routines of research, therapy and care, the authors in this collection bring together ethnographically-inflected historical and sociological observation with technically well-informed accounts of encounters between patients and professionals. They assess these in several national contexts in North America and Western Europe, and often through explicitly comparative analysis; the volume also juxtaposes different types of institutions, from community care to elite research-based facilities. Finally, in discussing different forms of malignant disease, the essays as a group reveal how patient populations and patient identities have differed. The picture that emerges is one of cancers rather than Cancer, of patients rather than 'The Patient', and of medical practices that are both experimental and routine.

This mix of the experimental and the everyday also makes studying cancer a particularly good tool for thinking about biomedicine and its ramifications more generally. Cancer is associated with especially high-tech innovations, but also seen as the 'Dread Disease', a threat that has brought the attention and funding which made those innovations possible. Furthermore, cancer's growing prevalence means that treatment for malignancies is becoming a common, even routine, experience. Thus the ways that cancer is managed – and the ways that cancer patients manage – have come to epitomise biomedicine. We hope, therefore, the chapters in this volume speak to broader questions about patients and their experiences within large institutions and with the sophisticated technologies of biomedicine, not only in the treatment of cancer.

We have structured the volume so that the essays in the first part emphasise individual trajectories, while those in the second part focus on the formation of pathways. The volume concludes with an essay reflecting upon the ways 'patient history' has been written, and suggesting an alternative approach to conceptualising the patient's role in the medical encounter.

## Patients

Part I of the book begins with Baines's account of how three generations of one family – her own – experienced breast cancer and its treatment. As an opening discussion, it sets the scene by examining personal patient experience, while also illustrating how new forms of treatment pathways have altered the patient's experience, not only of treatment but of the disease. Baines's account is informed by the illness narrative approach pioneered by scholars such as anthropologist Arthur Kleinman and sociologist Arthur Frank, and also by recent literature from the history of medicine dealing with breast cancer.[11] As she discusses how she, her mother, and her grandmother responded to the symptoms, diagnosis and treatment of breast cancer, she considers how they did or did not cultivate an identity as patient, cancer patient, or survivor. Her chapter also reflects on how cancer is experienced differently in routine treatment settings versus the research context.

The routinisation of treatment and its consequences for both patients and practitioners are also central to Timmermann's chapter (Chapter 3). By the 1960s, he argues, British lung cancer specialists believed that treatment had reached natural limits set by the cancer's biology, and thus that therapeutic gains would be marginal at best. Lung cancer, they concluded, could only be treated successfully in a minority of cases, and from the 1970s onward treatment for all patients followed a highly routinised pathway with a very limited range of clearly defined outcomes. For patients, though, the pathways through cancer treatment have never been routine. Drawing on an account written by a patient's wife in the 1990s, Timmermann illustrates how the patient underwent a series of diagnostic procedures, therapeutic decisions, and treatment modalities that blurred the distinction between curative and palliative care. At each point until very near the end, the patient and his family maintained hope of a cure and expected therapeutic success, even as it became increasingly unlikely. For the patient, the future was unknown; for the doctor it was all too clear.

Where Timmermann draws on a published personal narrative to explore the general contours of the patient's experience with routinised treatment pathways, Kutcher uses case records and personal papers to reconstruct one patient's experience of unknowingly being a trial subject undergoing highly experimental treatment. Maude Jacobs, an advanced breast cancer patient who had few treatment options left, underwent total body irradiation in the 1960s as part of research organised by Cincinnati's Dr Eugene Saenger, using cancer patients as proxies for soldiers exposed to nuclear warfare.[12] Kutcher considers Jacobs's fate from two different angles, as a sufferer of advanced breast cancer and as a subject in the trial of a radically new treatment modality. He also reveals how patients' stories can take on new lives, as when Jacobs's experience was reconceptualised yet again in the 1990s when ethicists and her family members looked back on her as a victim, of unethical experiments, exploitation, and a Cold War research agenda.

Barnes Johnstone also examines patient stories, but in terms of how they are used by patients, professionals and parents to guide young cancer sufferers through chemotherapy. Her account highlights how caring professionals have enlisted children with malignant disease, and their families, to create educational materials for other children. By drawing on their own interpretations of the rituals and tools of treatment, the child patients helped author comic books intended to make treatment look less frightening. Experts considered participation in the development of educational material to be therapeutic in itself, as this allowed some children a way to make sense of their experiences. Barnes Johnstone concludes that this participation also made space for child patients to exercise some previously unrecognised agency in their medical encounters.

## Pathways

While the essays in the first part of this volume emphasise patient experiences, those in the second part, 'Pathways', focus on how medical professionals and institutions have created and recreated the frameworks through which patients have experienced cancer treatment.

Löwy's chapter (Chapter 6) traces a debate among French and American surgeons and radiotherapists during the interwar and post-war decades regarding the ideal treatment pathway for early breast cancer. She shows that despite their agreement on the theoretical superiority of the radical mastectomy, many specialists dealt with women's resistance to this mutilating operation by offering them therapeutic alternatives. These doctors used various combinations of conservative surgery and radiotherapy, which was perceived as a less traumatic option, even as they argued that these alternatives were inferior as 'cures'. Löwy suggests that as uncertainty mounted regarding the prognoses for borderline lesions and biologically variable tumours, these doctors still felt it necessary to default to the most interventionist response they had available. Her international comparison reveals that in each context the notion prevailed that some responses to early breast cancer seemed obviously correct, the 'natural' pathway to take. But in practice, these 'natural' pathways differed, according to the local constellation of institutional and professional commitments.

The question of how to reckon with uncertainty also preoccupied the mid-twentieth-century British and American doctors who treated advanced breast cancer, as Toon's chapter (Chapter 7) discusses. These clinician-researchers conducted trials evaluating the efficacy of endocrine ablation and cytotoxic chemotherapy as treatments for women whose breast cancer had recurred after mastectomy. The serious side effects and variable responses to these systemic therapies made it difficult to discern whether or not they 'worked', and which patients were likely to benefit from them. To make sense of these confusing results, clinician-researchers created systems intended to render variable patient experience as objective data. But Toon suggests that this approach led them to focus their discussions on the aspects of the patient's trajectory that could be understood through quantification and statistical analysis, thereby pushing to the background elements of the therapeutic experience which did not lend themselves to such standardisation.

Keating and Cambrosio examine how the new style of practice based on clinical trials has reshaped the collective identities of cancer patients in the US and Europe.[13] From the 1970s onward, chemotherapy cooperative groups reorganised and extended clinical cancer research, seeking more trial subjects through community institutions and promising them better care in return. Keating and Cambrosio argue that a crucial tool in this transformation was the protocol, which allowed standardisation of treatment pathways and thus the comparability of therapeutic efficacy. Increasingly, group trajectories based on protocols have structured treatment pathways and experiences of cancer care, whether in a trial or not. The traditional view is that the growth of protocol-based pathways lessened patients' agency within clinical encounters: the protocol imposes a script on the patient, and on patients collectively, automating decisions that would otherwise come out of individual clinical encounters. But Keating and Cambrosio suggest that such a view of protocols is one-dimensional. The patients subjected to these new regimes were also an essential and sometimes scarce resource for trials, and so were able to exercise agency collectively: patient groups organised around specific diseases claimed to represent and speak for sufferers. Valier looks at the role of new screening and treatment technologies in the creation of treatment pathways for prostate cancer. She begins by describing how, in the 1990s US, prostate-specific antigen (PSA) testing increasingly detected abnormalities that might or might not manifest as full-blown cancer during the lifetime of the man screened; this in turn made prostate cancer a more visible and more aggressively-treated disease. Critics claimed that screening led to overtreatment, but advocates of early intervention argued that men's cancers merited as much attention as breast and other cancers for which women were screened. Meanwhile, as PSA screening increased the population of prostate cancer sufferers, proponents of a new and expensive treatment offered them a solution. Like total body irradiation in Kutcher's chapter, proton therapy was a product of Cold War military research then applied to the treatment of cancer, in this case for rare head and neck cancers. In the 1990s, promoters of proton therapy repurposed this modality for prostate cancer. Valier concludes by highlighting how market pressures together with patient activism and the 'preventive imperative' continue to subvert the logic of evidence-based medicine.

Up to this point, the chapters in this volume have analysed patient experiences and pathway formations retrospectively. Zuiderent-Jerak, Bal and Berg, by contrast, present a reflexive account of their own involvement in the institutional remodelling of cancer patients' trajectories. As ethnographically-oriented STS researchers, they evaluated care pathways in a Dutch university hospital, to help design efficient standardised trajectories that nevertheless offered patient-centred care. Their essay questions common sociological assumptions about standardisation, especially that the creation of standardised trajectories for patients is opposed to the provision of patient-centred care. Standardisation, they argue, is not necessarily inflexible and impervious to individual variation, and practices appearing to cater to an individual patient's needs might serve the interests of practitioners while impairing the experience of all patients. The interventions that most improve the patient experience individually and collectively are neither global nor radical, but grow out of evaluation of everyday work practices in particular settings.

Where Zuiderent-Jerak, Bal and Berg offer a reflexive account of pathway formation, Pickstone directs his reflexivity towards the practices of historians writing about patients. Why, he asks, did patient-focused histories began to appear in the early 1980s, and how did their emergence relate to national politics and local intellectual constellations? He draws on his experiences as member of an interdisciplinary network of scholars in Manchester – historians, sociologists, anthropologists, and public health workers – who were interested in understanding the social foundations of medicine in post-industrial Britain. Self-consciously 'social' historians of medicine then hoped to counter traditional practitioner-focused accounts by writing 'history from below'. But 30 years on, the rhetoric about patients – not only in the British National Health Service – is very different, as are the means of information and the conditions of history-writing. These shifts require us to revisit our conceptualisation of the patient-practitioner relationship both in the past and the present. Thinking of today's 'expert' patients as patrons, Pickstone argues, would help us do justice to the varied relationships constituting biomedicine by emphasising their political character, in a medical world that is more than just a marketplace.

## Notes

1 For an especially thorough overview of histories of cancer, see Cantor 2007. For more recent work, see Aronowitz 2007, Krueger 2008, Kutcher 2009, Löwy 2010, Keating and Cambrosio 2012.
2 Shimkin 1977, Laszlo 1995, Mukherjee 2010.
3 There are many scholarly works analysing these; for recent examples, see the essays collected from the Cancer Stories conference held at IUPUI (Indianapolis, IN) in November 2008, collected in the Fall 2009 issue of *Literature and Medicine*, especially the introduction (Schultz and Holmes 2009). On patients generally, see also Lerner 2006.
4 See Kleinman 1988, Frank 1995; and on the identity of the survivor: King 2006. The chapter by Baines in this volume shows how different generations in the same family do or do not accept the identity of 'survivor' or even of 'cancer patient'. Likewise, see Abel and Subramanian's (2008) examination of the women whose post-treatment syndrome leads them to adopt the 'wrong' kind of breast cancer identity, that of the person who remains ill rather than the survivor.
5 See for instance Lorde 1997, Diamond 1998, and Ehrenreich 2001.
6 For reproductions of and the background to Spence's work on cancer, medicine, and her experience of them, see Spence 1988 and Spence 1995; also see Bell's especially salient analyses of Spence's work (2002 and 2006) and discussions by her long-time collaborator, Dennett (2011 and 2001). Our great thanks to Terry Dennett, curator of the Jo Spence Memorial Archives, for the permission to reproduce this image.
7 Bell 2002, p. 17.
8 Ivan Illich (best known for his polemics against modern medicine along with other western institutions), did much the same, in refusing treatment for a tumour in a salivary gland diagnosed in 1983 – he died in 2002. See his obituary in the *Lancet*: Wright 2003.

9  See especially the chapters by Baines, Timmermann, and Kutcher in this volume.
10  For example Berg 1997, essays in Berg and Mol 1998, also Löwy 1997, Castel 2009.
11  On cancer (and illness) narratives broadly, see Kleinman 1988, Frank 1995, and Frank 2009; especially trenchant analyses of breast cancer narratives and accounts include Jain 2007, Segal 2007, Herndl 2006, Knopf-Newman 2004 and Potts 2000.
12  He explores this in more detail in his book: Kutcher 2009.
13  See also their recent book: Keating and Cambrosio 2012.

## Works Cited

Abel E. K. and Subramanian S. 2008, *After the Cure: The Untold Stories of Breast Cancer Survivors*, New York: New York University Press.

Aronowitz R. A. 2007, *Unnatural History: Breast Cancer and American Society*, Cambridge: Cambridge University Press.

Bell S. 2002, 'Photo Images: Jo Spence's Narratives of Living with Illness', *Health*, 6, 5–30.

Bell S. 2006, 'Living with Breast Cancer in Text and Image: Making Art to Make Sense', *Qualitative Research in Psychology*, 3, 31–44.

Berg M. 1997, *Rationalizing Medical Work: Decision-Support Techniques and Medical Practices*, Cambridge, Mass: MIT Press.

Berg M. and Mol A. (eds) 1998, *Differences in Medicine*, Durham, NC: Duke University Press.

Cantor D. 2007, 'Cancer Control and Prevention in the Twentieth Century', *Bulletin of the History of Medicine*, 81, 1–38.

Castel P. 2009, 'What's Behind a Guideline?: Authority, Competition and Collaboration in the French Oncology Sector', *Social Studies of Science*, 39, 743–64.

Dennett T. 2001, 'The Wounded Photographer: The Genesis of Jo Spence's Camera Therapy', *afterimage*, November/December, 26–7.

Dennett T. 2011, 'Jo Spence's Auto-Therapeutic Survival Strategies', *Health*, 15, 223–39.

Diamond J. 1998, *C: Because Cowards Get Cancer Too ...*, London: Vermillion.

Ehrenreich B. 2001, 'Welcome to Cancerland: A Mammogram Leads to a Cult of Pink Kitsch', *Harpers Magazine*, November, 43–53.

Frank A. W. 1995, *The Wounded Storyteller: Body, Illness, and Ethics*, Chicago: University of Chicago Press.

Frank A. W. 2009, 'Tricksters and Truthtellers: Narrating Illness in an Age of Authenticity and Appropriation', *Literature and Medicine*, 28, 185–99.

Herndl, D. P. 2006, 'Our Breasts, Our Selves: Identity, Community, and Ethics in Cancer Autobiographies', *Signs: Journal of Women in Culture and Society*, 32, 221–45.

Jain, S. L. 2007, 'Living in Prognosis: Toward and Elegiac Politics', *Representations*, 98, 77–92.

Keating P. and Cambrosio A. 2012, *Cancer on Trial: Oncology as a New Style of Practice*, Chicago: University of Chicago Press.

King S. 2006, *Pink Ribbons, Inc.: Breast Cancer and the Politics of Philanthropy*, Minneapolis: University of Minnesota Press.

Kleinman A. 1988, *The Illness Narratives: Suffering, Healing & the Human Condition*, New York: Basic Books.

Knopf-Newman, M. J. 2004, *Beyond Slash, Burn, and Poison: Transforming Breast Cancer Stories into Action*, New Brunswick: Rutgers University Press.

Krueger G. 2008, *Hope and Suffering: Children, Cancer, and the Paradox of Experimental Medicine*, Baltimore: Johns Hopkins University Press.

Kutcher G. 2009, *Contested Medicine: Cancer Research and the Military*, Chicago: University of Chicago Press.

Laszlo J. 1995, *The Cure of Childhood Leukemia: Into the Age of Miracles*, New Brunswick, NJ: Rutgers University Press.

Lerner, B. H. 2006, *When Illness Goes Public: Celebrity Patients and How We Look at Medicine*, Baltimore: Johns Hopkins University Press.

Lorde A. 1997, *The Cancer Journals: Special Edition* [orig. ed. 1980], San Francisco: Aunt Lute Books.

Löwy I. 1997, *Between Bench and Bedside: Science, Healing, and Interleukin-2 in a Cancer Ward*, Cambridge, MA: Harvard University Press.

Löwy I. 2010, *Preventive Strikes: Women, Precancer, and Prophylactic Surgery*, Baltimore: Johns Hopkins University Press.

Mukherjee S. 2010, *The Emperor of All Maladies: A Biography of Cancer*, New York: Scribner.

Potts L. K. 2000, 'Publishing the Personal: Autobiographical Narratives of Breast Cancer and the Self', in L. K. Potts (ed.) *Ideologies of Breast Cancer: Feminist Perspectives*, London: Macmillan, 98–127.

Schultz J. and Holmes M. S. 2009, 'Editor's Preface: Cancer Stories', *Literature and Medicine*, 28, xi–xv.

Segal J. Z. 2007, 'Breast Cancer Narratives as Public Rhetoric: Genre Itself and the Maintenance of Ignorance', *Linguistics and the Human Sciences*, 3, 3–23.

Shimkin M. 1977, *Contrary to Nature: Being an Illustrated Commentary on Some Persons and Events of Historical Importance in the Development of Knowledge Concerning Cancer*, Washington, DC: US Department of Health, Education and Welfare.

Spence J. 1988, *Putting Myself in the Picture: A Political, Personal and Photographic Autobiography*, Seattle: Real Comet Press.

Spence J. 1995, *Cultural Sniping: The Art of Transgression*, London and New York: Routledge.

Wright P. 2003, 'Ivan Illich', *Lancet*, 361, 185.

# Part I
# Patients

# 2

# Three Stories: Generations of Breast Cancer

*Joanna Baines*

## One

From an early age, I knew that my mother's mother had been glamorous. I had discovered old movies, and been captivated by the ladies of the silver screen: Lombard, Bergman, Colbert, and above all, Dietrich. Judging from my mother's photograph of her, my grandmother, with her Marcel-waved hair, elegant style, and enigmatic smile, would not have been embarrassed in such company. More evidence was provided by her house. Tastefully grand, it had two large copper beech trees on the front lawn, a tiger's head carved in the back of the fireplace, and a monster in the cellar, that growled whenever I went too close (though it was much quieter when my grandpa wasn't nearby). I also associated her with the smell of gas cooking. Although the association of a woman with her kitchen may be depressingly conventional, for me this also had an air of sophistication. Piped gas didn't reach the hamlet where I grew up until the late 1980s.

The definitive evidence of my grandmother's elegance and refinement, however, was the staircase. Consisting of two wide flights of stairs in the normal switchback arrangement, across the mid-way landing was another small flight of perhaps five or six steps, spreading the full width of both stairwells. Whether I had seen something similar in a film I'm not sure, but for whatever reason, this seemed the epitome of glamour, created just for the heroine, in a flowing silk gown, to gracefully descend, cigarette holder on high.

The only thing I actually knew about my grandmother was that she died of cancer.

This knowledge didn't destroy my romantic fantasies. Early Hollywood cancer, as Susan Lederer has described, was reliably clean.[1] There might be a paleness of the face, a little fatigue, perhaps a headache, maybe loss

13

*Figure 2.1*   The author's grandmother

of sight and, before anything unsightly could happen, a quiet, glamorous death, make-up maintained, lighting soft.

In 1966, my mother was 25 and had decided to enter nursing. After three months preliminary training at Piggotts Manor in Hertfordshire, followed by three months on the wards at Barts, she contracted glandular fever, a common hazard for nurses. Having been taken off duty immediately, she was sent home to Yorkshire to convalesce. She never returned. Instead, she stayed at home to nurse her dying mother.

My grandmother had suffered with 'indifferent' health for a long time, though there was little sign of this for years, other than, perhaps, her habitual intake of 'Bile Beans'. First manufactured in the 1890s, Bile Beans were advertised as promoting 'inner health and a lovely figure', and were a popular supplement in mid-century Britain.[2] Devised, allegedly, by a scientist impressed with the robust health of Australian Aborigines, they were recommended for a wide range of complaints, from constipation to rheumatism,[3] and were particularly useful for all those disorders 'peculiar' to women. At the height of their popularity, worldwide sales exceeded a million pills a day,[4] despite a 1905 court ruling in Edinburgh that declared the manufacturer's claims to be based on 'fraud' and 'impudence'.[5]

*Figure 2.2* The author's mother (top right) at Piggotts Manor

The exact nature of my grandmother's condition was never discussed with her children, although my mother is fairly sure that both her parents, towards the end at least, were aware that it was cancer, and terminal. Certainly, as she was finally taken in to the local infirmary, my grandmother acknowledged that she was seeing the surrounding countryside for the last time. As her condition had worsened, her youngest son had been sent to boarding school, and, after various examinations and procedures, including a barium meal, she had undergone an operation, described to the children as having her kidney 'capped'. My grandmother died at the age of 56.

The stage at which my grandmother was informed she was suffering from cancer is unknown. Certainly keeping the knowledge from patients for as long as possible was a common practice, although by mid-century, non-disclosure was increasingly being questioned. In the US, following a lively discussion on the non-disclosure at the New York Academy of Medicine, a book was published in 1955 entitled *Should the Patient Know the Truth? A Response of Physicians, Nurses, Clergymen, and Lawyers.*[6] As the title suggests, the book's goal was to address the issue from a wide variety of perspectives, the notable exception being that of the patient. Although never intended to address any specific condition, cancer was the dominant theme. Of the 23 chapters, six mentioned no specific conditions, nine spoke of cancer only, and five discussed cancer and other conditions, with cancer often still the most prominent and most problematic issue. The remaining three chapters, those that mentioned conditions other than cancer only, addressed the specialty areas of mental health care, childbirth, and cardiology. The key concern for all the professionals, as in the US in general, was the maintenance of hope,[7] and although opinions were mixed, most contributors leant towards a case-by-case assessment, relying on the physician's discretion.

The disclosure issue was also debated in Britain, where the medical profession was traditionally more paternalistic, and the public, allegedly, was averse to 'fussing over cancer'.[8] In 1959, doctors, clergymen and other interested parties debated the subject in the letters page of the *Times*. Mr A. Dickson-Wright, vice-president of the Royal College of Surgeons and treasurer of the Imperial Cancer Research Fund, had given a lecture at the London Rotary Club. A brief newspaper report included his claim that cancer patients should never be informed of their condition, even when cured. This provoked numerous responses in the days that followed, many questioning whether any physician had the right to keep the truth from a patient who may wish to put their affairs in order, and pointing out the dangers that might result from a lack of trust between doctor and patient.[9] Some letters, including one from Robert Platt, president of the Royal College of Physicians, took the middle line, believing that each case was individual, and that doctors had to balance the need to retain hope, with the need to avoid dishonesty.[10] Two correspondents, both doctors, firmly supported Dickson-Wright. A general practitioner in Bristol stated he never informed his own patients, but had met two people who had, at their own insistence, been told the truth:

> Never have I seen such utter dejection, such lack of desire to go on living, and such lack of fighting spirit, which quite often carries a patient through for periods far longer than a doctor could, at his most optimistic, expect.[11]

At a prestigious lecture the following month, Dickson-Wright defended his statement. To inform a man he was about to die was 'impertinent', and the increasing calls for doctors to be 'cruel and factual' should be resisted. These calls were particularly strong in America, where the recent death of the Secretary of State, John Foster Dulles, had generated the 'most disturbing publicity'. Dickson-Wright, along with many others in Britain, remained unconvinced that cancer education could result in anything but cancerphobia and concealment, the currently improving survival rates having done nothing to diminish the terrible spectre of the disease. Defending his position, Dickson-Wright described cancer as 'our great remaining uncertainty in the world of medicine'.[12]

In the same year a survey of the opinions of cancer patients was conducted by the Manchester Committee on Cancer. Only 7 per cent of those asked disapproved of having been told of their cancer diagnosis, although 19 per cent denied they ever had been told, despite evidence to the contrary. Many of these patients believed that knowing the truth had helped them to fight the disease more effectively.[13]

## Two

In 1976, when I was seven and my mother was 35, she heard a radio programme on breast self-examination. Having found a lump, she was sent to have a biopsy. Shortly afterwards the local doctor phoned to inform my mother that the result was positive, and 'the breast will have to come off'. She underwent a radical mastectomy and was kept in hospital for a week to allow the wound to drain. She has very few memories of this time, apart from of a few other patients, particularly one old lady, evidently dying and continuously calling for her mother, while her family sat by helplessly. It was an old hospital, gloomy and dingy, with round windows on the doors, and reminded me of the ships and submarines my father, a design manager at the local shipyard, regularly received tickets for us to visit.

After draining, my mother was sent to a convalescent home for a fortnight. There were patients of all ages, the women having all been treated for cancer, while the men had not. Visiting her I always tried to take something nice – one time a handful of bright, shiny bottle tops from my father's home brewing kit; another time, a bag of conkers, which my mother would later drop noisily down a flight of stairs, infuriating Matron. I had very little idea of what was going on at the time. The older girls at school surrounded me one playtime. They'd heard my mother was ill, and asked what was wrong with her. I couldn't remember, but thought it started with a 'c', to which they suggested cancer, and I nodded enthusiastically, pleased to be able to answer them. They were uncommonly nice to me for some time afterwards. My mother was released on Christmas Eve, going straight from the convalescent home to queue in the butchers' for our Christmas meat.

I don't remember seeing my mother's chest for the first time. She didn't hide it from me, as she hadn't before, and I have no recollection of shock or disgust, more curiosity. By the time she underwent her radical mastectomy, the procedure had been losing favour for some time, the 1970s being a period when, with the support of the feminist and patient rights movements, women were beginning to demand both a role in medical decision-making and less mutilating treatment options.[14] Criticism of the procedure itself, however, had a far longer history.

The man whose name is forever linked with the operation, William Stewart Halsted, was a surgeon at Johns Hopkins University Hospital. In a paper delivered to the American Surgical Association (ASA) in 1907, he described what he claimed were impressive success rates for the 'modern, so-called complete, operation' at Johns Hopkins.[15] The fact that American surgeons in the generation preceding him admitted to never curing a case of breast cancer, demonstrated to Halsted that the disease, as then recognised, had already entered the lymphatic vessels.[16] Therefore, the operation was necessarily extensive:

> ...we must remove not only a very large amount of skin and a much larger area of subcutaneous fat and fascia, but also strip the sheaths from the upper part of the rectus, the serratus magnus, the subscapularis, and at times from parts of the latissimus dorsi and the teres major. Both pectoral muscles are, of course, removed.
>
> A part of the chest wall should, I believe, be excised in certain cases, the surgeon bearing in mind always that he is dealing with lymphatic and not blood metastases.[17]

Halsted's justification for such widespread intervention was his belief that the disease spread in an orderly, centrifugal manner.[18] This belief was shared by Charles P. Childe, surgeon at the Royal Portsmouth Hospital who, in the same year that Halsted addressed the ASA, published his treatise, *The Control of a Scourge or How Cancer is Curable*, a book that would play a major role in inspiring cancer education initiatives on both sides of the Atlantic.[19] Halsted and Childe both cited as evidence the microscopic analyses of W. Sampson Handley at the Middlesex Hospital, who had demonstrated that the number of cancer cells present in body tissue declined the further away from the primary tumour that tissue was situated.[20] Childe also presented as confirmation a plaster cast of a female breast cancer victim in St Thomas's Hospital Museum, which displayed secondary cancer in every bone apart from those below the knees and elbows.[21] Building on this logic, a number of surgeons introduced the super radical mastectomy, a procedure in which the clavicle was divided and the first rib resected, allowing even more of the patient to be removed. This procedure, however, never became popular among surgeons.[22]

Despite this long history, there has rarely been a time when the radical mastectomy went unchallenged. One of the earliest and staunchest critics of the procedure was Geoffrey Keynes at Barts, where the Surgical Professorial Unit was awarded a supply of radium in 1922.[23] Having demonstrated that a recurrent carcinoma could be destroyed by local radiotherapy, the process was successfully applied to patients with early stage breast cancer, who would formerly have undergone radical surgery.[24] In a letter to *The Lancet* in 1929, Keynes bemoaned the fact that the 'severe and mutilating operation' based on the work of Halsted and Sampson Handley, had been practically 'imposed' upon surgeons as the 'standardised' treatment. Praising the successes of radiotherapy, Keynes believed it unlikely he would ever perform the radical procedure again.[25] A surgeon who strongly opposed much routine surgery, including the common practice of circumcising newborns apparently in distress when passing urine,[26] Keynes was respectfully tactful when addressing the ASA in 1937, in a country where allegiances to the radical procedure were always stronger than in Europe,[27] referring to Halsted as their 'great countryman'.[28] Nearly half a century after Halsted had first introduced the radical mastectomy, Keynes expressed his 'increasing difficulty' in accepting Sampson Handley's theory of 'centrifugal permeation',[29] and extolled the virtues of radium.[30]

Another critic was Robert McWhirter of the University of Edinburgh,[31] who in 1948 claimed that the high success rates attributed to the radical mastectomy were a result of 'careful selection', restricting the procedure to early, favourable cases.[32] Simple mastectomy (removal of the breast tissue only), and the speedy application of radium to both the wound (which healed considerably more quickly following the more conservative operation), and the axilla, not only could result in improved survival rates, but would not run the risk of 'edema' of the arm.[33] Now commonly referred to as lymphoedema, this swelling, caused by the accumulation of lymph fluid, can result when the axillary nodes have been removed. McWhirter's faith that a simple mastectomy with 'roentgenotherapy' was preferable to the radical procedure was not without caveat however. Radical mastectomy was preferable if the patient was 'stout', it apparently being difficult to deliver an adequate dose of radium in such cases.[34]

McWhirter's approach was challenged in 1958 by results from the Mayo Clinic in Rochester, which appeared to demonstrate higher success rates for the radical procedure.[35] However, as Cleveland surgeon George Crile Jr, son of the prominent surgeon, George Washington Crile, and a long-time proponent of more conservative measures pointed out, once the advanced stage of many of McWhirter's patients was taken into account, the difference was negligible.[36] These 'winds of controversy' from Edinburgh and Minnesota, as *The Lancet* poetically described them, did little to 'dispel the mist' surrounding the issue at the time.[37] In 1972, Crile Jr, who published articles and books on the subject for both the public and the profession, presented the results of a longstanding trial comparing survival rates between mastectomy and local excision. Results were similar. Crile concluded by discussing how strong the fear of mastectomy had been in some of the patients, and suggested the availability of less mutilating operations might well induce women to present for treatment earlier.[38] In Crile's support, an analysis of the Johns Hopkins Hospital records between 1889 and 1931 demonstrated that, despite Halsted and his pupils reducing operative mortality and local recurrence, their efforts had made no impact on long-term survival.[39] Gradually, as medical evidence of the effectiveness of conservative surgery combined with radiotherapy increased, along with women's awareness of their options, and their willingness to fight for them, use of the radical mastectomy slowly declined.

A *Lancet* tribute to Halsted in 1952, the centenary of his birth, listed the areas in which he had made significant contributions as wound infection, vascular surgery, intestinal anastomosis, surgery of inguinal hernia, goitre and breast carcinoma, and anaesthesia, during which work he acquired and overcame an addiction to cocaine. Halsted had also been responsible for the introduction of rubber gloves to the operating theatre in the late nineteenth century. Having noticed that contact with a surgical solution was causing a skin inflammation on the hands of Caroline Hampton, theatre nurse and future Mrs Halsted, he asked the Goodyear Rubber Company to create a suitable pair of gloves.[40] The *Lancet* tribute concluded by praising Halsted's unhurried, gentle and meticulous surgical technique, along with his belief, constantly drummed into his students, that surgery had to be for the good of the patient, not the glory of the surgeon.[41] Unfortunate then perhaps, for his legacy that Halsted should now be seen by many as the epitome of that contemptible band, the 'heroic' surgeons, and their arrogant, gung-ho pursuit of accolades.

Following her operation, my mother underwent ovarian ablation by radiotherapy, for which she was admitted for three days to Christie Hospital, Manchester. The impossibility of more children was discussed with her extremely seriously. That she would enter instant menopause was not, although, as she might suffer a few hot flushes and 'some discomfort', she was offered hormone-replacement therapy (HRT), which she declined. That there might be a hormonal factor in the causation of breast cancer had long been suspected. George Beatson, at the Glasgow Cancer Hospital, was the first to publish on surgical removal of the ovaries, 'the seat of the exciting cause', in 1896, as a palliative treatment for advanced breast cancer.[42] Radiotherapy would replace surgical removal by the early twentieth century, although with radiotherapeutic techniques being initially less effective, the practice soon fell out of favour.[43] At times referred to as 'castration',[44] the practice was initially assessed in the first randomised control trial in cancer treatment, carried out at Christie Hospital, Manchester, between 1948 and 1955.[45] Despite this and subsequent studies often showing the process to be at least slightly advantageous, it was not until the systematic review by the Early Breast Cancer Trialists' Collaborative group, in the 1990s, that the practice could be said to have been fully endorsed.[46] A 1963 reply in the *British Medical Journal*'s 'Any Questions?' section illustrates both this uncertainty, and the rampant paternalism of the British medical profession at the time:

> the degree of protection afforded is not very considerable…a woman might well prefer to run the slightly increased risk of developing metastases rather than undergo such an operation with its considerable physiological and psychological consequences. It is rarely possible to discuss this problem frankly with the patient, nor is it probably fair to discuss it with the patient's husband. In most instances the decision should be made by consultation between the surgeon and the family doctor.[47]

Given this connection, the possibility that HRT might cause or exacerbate breast cancer has been debated for decades.[48] Both the Women's Health Initiative trial and the Million Women Study demonstrated in the early years of the new millennium that HRT increased the risk of breast cancer (among many other conditions) in healthy women, and on 17 December 2004, the HABITS (hormonal replacement therapy after breast cancer – is it safe?) randomised trial was terminated, when early results showed an unacceptable level of recurrence.[49]

On attending a prosthesis clinic, my mother was given a fake breast with a plastic front and soft back. Resembling a deformed fried egg, this began to shrivel up almost immediately, and six months later, was replaced by a soft cloth version, ordered through a magazine advertisement. Now, she rarely wears a prosthesis at all. My mother was called in for follow-up examinations at increasing intervals over the five years following her surgery, during which her remaining breast and neck glands were the sole focus of attention. Once the five years were over, she was told there was no need for any more check-ups, unless she had particular concerns. She was, however, entered into a clinical trial, involving a yearly mammogram. The only information about her condition my mother was ever given by her physicians was that the treatment she received was necessary. There was no reading material, no access to counsellors or support groups, and, although she was told to carry on checking her remaining breast, there was no mention of the possibility of secondary metastases. My mother decided almost immediately that the doctors had got it wrong. She didn't have cancer. The quicker she went along with what they wanted, however, the quicker life could get back to normal. She found the convalescence in the company of other patients not only useful, but fun. The main and enduring annoyance was the inconvenience of the prosthesis, particularly when trying to find clothes. After Christmas, with me back at school and my father back at work, she did become depressed, but only, she reasons, 'as much as someone would after any operation'. Later she felt the experience had enabled her to worry less about everyday problems, and make a bigger effort to enjoy life, but she is unsure of the timeframe involved. My mother finds the idea of being a 'survivor' laughable, and still does not use the word 'cancer'. If the subject arises, she will state 'I had a mastectomy'.

## Three

Such denial was never an option for me. What had been a short-term, fairly liminal experience for my mother had, by the time it came to my turn, evolved into a lengthy, highly detailed and all-encompassing process. I was never advised that a possible genetic factor meant I should be extra vigilant, though in my teens, I happened to watch a television programme that reported family history as a risk factor for cancer. The programme also, helpfully, proclaimed worry another factor. I found a lump in the shower, when I was 32 and a mature student in the first week of exams at the end of my anthropology degree in London. Had I not already had a doctor's appointment scheduled for two days later (to discuss stress-related anxiety and possible depression) it's doubtful I would have acted as quickly as I did. I raised the issue of the lump at the end of the session, practically as rising to leave. The doctor thought it felt fine but that, given my family history, I should be referred. It took my mind off the few small exams, projects and dissertation I still had to finish.

The surgeon I was referred to also thought the lump felt safe, but sent me for an ultrasound, a machine I had only ever seen on television, giving expectant parents a first glimpse of their offspring. I suppressed the urge to coo when the small mass appeared on the screen. No heartbeat. Again, my surgeon emphasised that it was probably fine, but I should return for a fine-needle aspiration biopsy, just to be sure. At the biopsy, the ultrasonographer and nurse were both female, much older than me, and, having succeeded in retrieving a part of the tumour, decided I deserved an ice-cream, having stoically endured their heroic struggle. Afterwards, the ultrasonographer sat down beside me in the waiting room and informed me it was their job to worry about the lump from now on. I should leave the worrying to them.

When I went in for the results I hadn't noticed that the day of my appointment had changed to Wednesday, the others having all been on a Monday. It was only much later that I realised the significance of this, that I had already made the switch, from a possible cancer patient to a confirmed one. When the breast care nurse came out to send me for a mammogram, something I had earlier been told was practically useless in someone of my age, I demanded to know whether I had cancer or not. She calmed me down, and told me we would discuss it after the mammogram. In the consulting room, I ignored the illuminated mammogram films, which my surgeon examined with a vaguely confused look, as if mystified as to why this should be happening to me. I concentrated instead on the floor or the window. I had always had slight difficulty breathing in hospitals, and would spend most of the very small amount of time I had spent in them checking there was a window open, or, failing that, an air vent or fan, anything that demonstrated air was moving. Years later I connected this feeling with my breech birth ('Awkward from the beginning!'): when silent and blue, I had been quickly whisked off to an incubator. I'm not entirely convinced this was the cause, but it has a certain logic. By the time I made the connection, however, I had spent so much time in hospitals that the feeling had gone. Cancer had cured me.

Having delayed the surgery by a week, I met my breast care nurse to go over everything again, then went into hospital for a lumpectomy and sentinel node biopsy, where only a few of the axillary nodes are removed for testing. In the bed next to me was my Fellow Traveller. She had been in the mammogram waiting room on the day I was diagnosed, and overheard me panicking about the possibility of having breast surgery. Quietly, she had leant forward and told me that it wasn't as bad as I was imagining – she'd had many operations on her breasts. Now she was here again to keep me sane, reassuring me that my snoring was very gentle the day it woke me up, and laughing as we compared her drain to a dog, dragging along behind her, unnamed so she wouldn't get too attached. Across the room was a stern-looking, older lady, whose husband came regularly, to sit quietly on her bed while she talked. My visitors named them after one of our more forthright friends and her quieter fiancé. For my entire three days as an in-patient, I was terrified of using the toilet, as this lady had little warning when she needed to go. Twice, when too groggy to venture further, I went in, only to have her pounding on the door. Each time I emerged too late, and had to find a nurse to clean up. One day, when visited by what we assumed were her children, her voice rang out even louder than usual: 'The only good Palestinian is a dead Palestinian!' I saw her again many months later, entering a lift at Charing Cross Hospital. She was older, shorter, somehow less defined, and leaning heavily on her husband. Having noticed me, her face became even more confused as she tried to remember how she knew me. Then the lift doors closed.

Being bored lying in bed, I would go out as much as possible. Despite always asking if this was alright, I would still be reprimanded, had a doctor been looking for me. One time I arrived back to hear a rather uptight, young, female doctor, bemoaning the fact that 'Mrs' Baines was never here, and where on earth did I go? On the morning of my operation, a young man came round to draw a large black arrow, pointing to the breast in question. This was not particularly reassuring. I was obliged to go in a wheelchair for my pre-operation ultrasound, where I was handed my notes and given some vague instructions on how to find my own way back to the ward.

The lumpectomy wasn't enough, no clear margin. My surgeon (the best, people kept telling me, though I doubt anyone was told they had the second or third best) wanted to do a mastectomy, as (he said) my breasts were small. This was quite a surprise as I was a G cup at the time, so I refused, until eventually he examined me and agreed my breast was certainly large enough for another attempt at a lumpectomy. A man who had been introduced to me, but whose name and purpose I hadn't registered, rocked back and forwards with his hands in his pockets, and added yes, my surgeon had certainly done me a disservice! Was this really not as creepy as it now seems, or was I so thankful at the time I didn't care? As part of the negotiation I agreed to an axillary clearance (although the selected nodes had been fine), therefore increasing my chance of lymphoedema.

No Fellow Traveller this time. In one of the long, long nights, a young woman in the bed across from me began to make strange noises.

Her neighbour and I looked at each other, unsure whether she was crying or laughing (or dying!). It was relieved laughter. She had eventually managed to break wind, a 'good sign' apparently, and we congratulated her. I was given two drains this time, which fitted nicely in my dressing-gown pockets, although as the pain was far more noticeable this time, my wanderings were rather restricted. I still managed to dislodge my drains on a number of occasions.

Just over a week after I came out, two non-driving friends and I were moving out of the university halls into a nearby flat. A draining session was therefore scheduled for the morning I would transport our worldly goods in a hired transit. A few days after moving in, I met another doctor to arrange my tests. First was a CT or CAT (Computerised Axial Tomography) scan on my abdomen. To enhance the results, a dye was injected into my vein, which I could feel moving its way down from my ears. And then I wet myself. In my horrified panic, I forgot I could talk to the radiographer in the control room, and just flailed an arm around helplessly. She calmly explained that the feeling was just a common, but misleading side-effect of the dye, and however convinced I was otherwise, nothing had happened. Not something they could have warned me about then.

Next on the list was a liver ultrasound. The technician moved his screen around, occasionally out of my sight. Was he hiding something ominous? Would I recognise it, even if I could see the screen? Then a bone scan in the Nuclear Medicine Department, presumably so-named because no-one could come up with a scarier title. I received two information sheets, one a list of questions and answers. Was I a hazard to other people? No. The other sheet advised me to keep contact with pregnant women and small children to a minimum for the rest of the day.

A couple of weeks later I went to Charing Cross Hospital (which isn't at Charing Cross) to meet my chemotherapy Professor, who informed me I might not need any. The next week, without further explanation, he told me I should. I didn't ask. So, four months after finding the lump, I began chemotherapy. Throughout that time I'd been told I would probably receive the CMF combination (cyclophosphamide, methotrexate and fluorouracil), but the treatment evolution was continuous, and now it was to be fluorouracil, epirubicin, and cyclophosphamide, the gloriously-named FEC regime. For the first session I wore a head cooler in a doomed attempt to save my hair. I quickly reconciled myself to a (strangely satisfying) head shaving routine, and went for a pirate look; hooped earrings and a bandana.

Chemo was every three weeks for nearly five months. As the first dose was a breeze, my steroid intake was dropped for the second round. Given the human body's remarkable capacity to forget the bad times, I can't really remember how I felt that night, only that I knew I couldn't do this again. My steroid levels were re-elevated, and the remaining doses were comparatively uneventful. Boarding the 220 bus afterwards, I would quickly sit down wherever I could avoid the eyes of the old people, glaring balefully at anyone they thought should relinquish their seat. I reassured myself grimly that I'd probably be dead before any of them.

The chemo suite was on the sixth of the 15 floors at Charing Cross. Close by, with prominently displayed hand-sanitiser dispensers on sentry duty, was an in-patient ward, with subdued lighting and a tunnel-like entrance. Occasionally a bed-ridden patient would be wheeled out to the lifts, where we, the walking wounded and our associates waiting at the regular-sized lifts, would studiously examine our watches or feet, willing the lift to hurry.

December came, and time to plan Christmas back up north. Having a chemo session scheduled for 23 December, and given there would be quite enough forced gaiety to endure as it was, I decided to go home earlier, and therefore avoid the event itself. A few days before I was due to leave, however, the hospital rang. My last blood tests had shown my white blood count was dangerously low, and I should avoid contact with anyone. Having refused my flatmates' kind but reluctant offers to stay with me, for Christmas lunch I made myself a fry-up. For that year it was the perfect Christmas.

My blood count collapsed again later. I was recruited into the trial of a drug designed to increase appetite, held at Hammersmith Hospital (which isn't in Hammersmith), and for which I would be paid £100. After eating my carefully weighed cereal, I settled down to watch *Gosford Park* when my blood results came through. I was immediately put in a taxi and taken to Charing Cross, where two doctors I hadn't met before asked me if I felt okay, along with countless variations on that theme, while looking as though they expected me to keel over dead at any moment. I was taught how to inject a white blood cell boosting substance into my stomach. Four days later my blood count was back up, but it was decided I'd had enough chemo, I could miss the last session and go straight to radiotherapy. Lengthy preparations: the creation of a mould, calibration of machines, and application of minute tattoos, then a short burst, every-day for two weeks. I didn't realise when people warned me how tired I would get, that they weren't even slightly exaggerating. I was also put on tamoxifen for five years.

Tamoxifen, initially known as ICI 46,474, was first synthesised by Dora Richardson in the ICI laboratories at Alderley Park, just outside

Manchester, during attempts to find a 'morning-after' contraceptive. Arthur L. Walpole, then head of the fertility control programme, encouraged staff at the Christie Hospital to carry out clinical studies on ICI 46,474, and in 1973, Nolvadex, the ICI brand of tamoxifen was approved for the treatment of advanced breast cancer in the UK. Since then, tamoxifen has become the 'gold standard' of early breast cancer treatment.[50]

I was also asked if I would take part in the HERA trial of Herceptin in the treatment of early-stage breast cancer. Having been persuaded at great length that the drug would probably be a good thing for me, I was then allocated to the non-treatment leg. Every three months I was given a MUGA (Multiple Gated Acquisition) scan, to ensure my heart wasn't being affected by the drug I wasn't taking. Having undergone many of these tests, I was told, without any reason offered, that the MUGA didn't work very well on me, and given echocardiograms instead. Two years later, the interim results were so impressive that everyone in the trial was offered the chance of having treatment. While the press railed at the postcode lottery of Herceptin provision, with women offered or denied the drug dependent on where they lived, I began two years of treatment.[51]

Shortly after finishing Herceptin (which, apart from weakening my nails, had no discernible side effects), I realised my five years on tamoxifen were coming to an end. Having mentioned this at a check-up, I was informed that trials were ongoing into whether five or ten years of treatment was best, but as yet, results were inconclusive. We all live (and die) on the cutting edge.

## Pathways

I have no idea what information my grandmother was given, at what stage she was told of her diagnosis and prognosis, or the specifics of any treatment she received, and have no real idea what sense she made of her situation, in a world reluctant to use the word 'cancer', let alone 'breast'. However, from these brief descriptions it is clear that my mother and I not only experienced different disease regimes, we experienced different diseases. Removed from normal life, and returned when apparently recovered, the total lack of support services or room for treatment negotiation made it clear she should not 'make a fuss', while with examinations largely restricted to her breast area, and the possibility of secondary tumours unacknowledged, she was presented with a localist understanding of the disease. For my mother, cancer was an event, not a process: acute, not chronic: local, not systemic.

At the time of my diagnosis it was immediately made clear to me that life had changed, with active treatment for up to a year, and medication and extensive check-ups for as long as I could imagine. Copious amounts of information and invasive medical investigations established cancer as a part of me and my life, whether tumours, the physical manifestations, were present or not. The overall message conveyed was that cancer, even if advanced was manageable, if only to a limited extent, and I would play a major role in that management.

*Figure 2.3*   Two generations

## Meaning

Every cancer patient has to ascribe meaning to their disease, to incorporate it in some way, into the narrative of their life. For my mother cancer was an inconvenience, an obstacle to be overcome and consigned to the past. For her it was, additionally, a mistake. Having been the closest witness of her mother's death, this was perhaps a necessary coping strategy, one made possible by the liminal nature of her treatment and, of course, the absence of secondaries. As the temporal range of cancer is extended virtually indefinitely, and the number of people living in the 'remission society' multiplies,[52] it is increasingly impractical for patients to view their cancer in this way. Given this 'new' cancer, increasingly a part of life, and of personhood, it is hardly surprising that the need for the experience to constitute a positive contribution to life is also greater than ever.

Having received my diagnosis, my response was to research the problem. This is, of course, what any 'responsible citizen' (and certainly someone with the resources of an entire university to hand) should do. Now, however, given that I spent much of those early days scaring myself with statistics (five years survival = 81 per cent, ten years = 68 per cent, fifteen years = don't ask), I suspect it had more to do with making the situation seem real, in a time before the sheer length and all-encompassing nature of treatment had made the reality inescapable.

Having exhausted myself on the cold, hard facts of science, I turned to the 'crowded market'[53] of cancer memoirs. Scholars have frequently raised the issue in recent years of whether it is possible to respect the right of these cancer narrators to make the choices they have made, and yet engage critically with the results.[54] When the narrators in question presume, either explicitly or implicitly, to dictate how one should be ill, and, as Judy Segal has demonstrated in the case of breast cancer, to maintain ignorance surrounding the production of information, therefore effectively closing down dissent and obscuring alternatives, then there has to at least be debate. What is missing from the generic account of the individual survivor, bravely waging war on her disease?[55]

The objective of illness narratives was beautifully described by one of the earliest people to consider them worthy of exploration, Arthur Kleinman, in a sentence that encapsulates both the promising, and problematical, aspects of the genre:

> The point I am making is that the meanings of chronic illness are created by the sick person and his or her circle to make over a wild, disordered *natural* occurrence into a more or less domesticated, mythologised, ritually controlled, therefore *cultural* experience.[56]

Nowhere has this ritualised mythology become more dogmatic than in the breast cancer narrative, where the weight of history, of the women who have struggled for acknowledgement, for choice, for meaning, and for life, renders any criticism seemingly downright disrespectful. But then I doubt many of the early pioneers would be so circumspect. I have difficulty imagining Audre Lorde submitting to the moral oppression of the 'think positive' mantra, and the total absence of discussion, in Britain at least, of possible environmental factors, in favour of the individualistic, 'I will survive' attitude, would have Rachel Carson in abject despair.[57]

Arthur Frank, a prolific commentator on illness narratives, has concentrated particularly on the 'rhetoric of self-change', whereby the author credits their illness with initiating a positive transformation of their life:

> This ethics takes its place in a genealogy that runs from early Christian confessional practices through psychoanalysis to 12-step groups and other forms of contemporary self-help such as 'journaling' groups; it is an ethics of telling the truth of oneself on the premise that truth can be found in the telling. The promise of this ethics is that telling certain confessional truths will somehow purify (redeem, heal, improve, reconcile, detoxify, etc.) both the teller and others. Truth *needs* to be told.[58]

In the rhetoric of self-change, illness is seen as prompting an 'epiphany', revealing to the patient what is wrong in their lives, therefore giving them the opportunity to change, and, like a 'phoenix', rise again.[59] Within Frank's terminology, I am probably a reluctant phoenix.[60] Certainly, I could never deny my experiences of cancer any redeeming qualities: to do so would be tantamount to invalidating the last ten years of my life (and even, perhaps, however much time I still have to come). It could beargued that I have gained a greater understanding of anxiety, a condition I have been subject to, off and on, for much of my life:

> Anxiety, like many other mental problems, is an excess of meaning, and an inability to escape from that excess, a condition the 'cancer experience' affords countless opportunities for development: straining to hear comments as the nurses organise notes and prepare to call a patient in; analysing every facial expression, body movement, and word spoken (or unspoken) by the doctors; secretly examining x-ray films, despite having no idea what might be ominous, what benign; and of course, the constant interpretation of the everyday signs and symptoms of an embodied life. The only time in the past eight years that I have not been on medication for anxiety, I became unable to sleep for longer than two hours, was constantly on the verge of tears, and left my consultant with no option but to refer me for an invasive, painful, and unnecessary procedure.

If this is epiphany, give me ignorance.

Along with many commentators, I share the belief that illness narratives should, and could, accommodate chaos, but am at a loss as to how this might actually be achieved.[61] All narrative structuring, whether in mind or on paper, will serve the instigator's needs at the time, but when the master narrative is so overwhelming, as it is with breast cancer, the canon well-nigh hagiographical, how much freedom does the narrator have? Or, following Frank, in Foucauldian terms, when does care of the self become a technology of the self?[62] I also share Shlomith Rimmon-Kenan's concern that narrative can lead to what she terms 'textual neurosis',[63] rendering events sterile and obsolete, not only neutering what is being conveyed, but also naturalising the composition for the narrator. All recollection is selective. The problem is, once formed, it may become directive.

Why can't they ever let my wanderings alone?! Can't they understand that I'll talk it all to pieces if I have to tell about it? Then it's gone, and when I try to remember what it really was like, I remember only my own story. – Snufkin in *The Spring Tune*.[64]

## Notes

1   Lederer 2007.
2   Rowe 2003.
3   Rowe 2003, p. 138.
4   Rowe 2003, p. 139.
5   Rowe 2003, p. 138.
6   Standard and Nathan 1955.
7   Cantor 2006.
8   Toon 2007, pp. 120–2 and 137; Cantor 2007, pp. 9–10; Patterson 1991, p. 142; quotation from Hillman 1928, p. 479.
9   Jankins 1959; Adams 1959; Rouse 1959; Sinclair 1959; Nicolas 1959.
10  Platt 1959; Howard S. 1959; Pantin 1959; Francis 1959.
11  Howard H. 1959.
12  Toon 2007, pp. 124–8; Anonymous 1959a. On the effect of John Foster Dulles' cancer see Lerner 2006, pp. 81–99.
13  Aitken-Swan and Easson 1959.
14  Lerner 2000, p. 27.
15  Halsted 1907, p. 2.
16  Halsted 1907, p. 4.
17  Halsted 1907, p. 16.
18  Halsted 1907, pp. 9–10.
19  Aronowitz 2001, p. 356; Toon 2007, p. 119.
20  Handley 1905, pp. 983–6; Childe 1907, p. 54; Halsted 1907, p. 9.
21  Childe 1907, pp. 27–8.
22  Robinson 1986; Lerner 2001, p. 82.
23  Lerner 2000, pp. 35–6.
24  Keynes 1928, p. 108.
25  Keynes 1929, pp. 156–7.
26  Keynes 1930, pp. 52–3.

27   Lerner 2000, p. 38.
28   Keynes 1937, p. 619.
29   Keynes 1937, p. 620.
30   Keynes 1937, p. 622.
31   Robinson 1986, p. 329.
32   McWhirter 1949, p. 830.
33   McWhirter 1949, p. 834.
34   McWhirter 1949, p. 840.
35   Anonymous 1958, pp. 249–50.
36   Crile 1958, p. 739.
37   Anonymous 1959b, pp. 1186–7.
38   Crile 1972, p. 550.
39   Baum and Edwards 1972.
40   Lerner 2001, p. 18.
41   Anonymous 1952.
42   Clarke 1998, p. 1246; Beatson 1896, p. 163.
43   Robinson 1986, p. 326.
44   Nissen-Meyer 1965, p. 249; Paterson and Russell 1959, p. 130.
45   Paterson and Russell 1959, pp. 130–3; Clarke 1998, p. 1247.
46   Clarke 1998, p. 1248.
47   Anonymous 1963, p. 1390.
48   Anonymous 2004, p. 410.
49   Holmberg and Anderson 2004, pp. 453–5; Anonymous 2004, p. 410.
50   Jordan 1998.
51   Anonymous 2005, p. 1673.
52   Frank 1991, pp. 138–9.
53   Rabinovitch 2007, p. 195.
54   Broom 2001, pp. 249–50; Herndl 2006, p. 235.
55   Segal 2007.
56   Kleinman 1988, p. 48.
57   For a particularly elegant dissection of 'thinking positive' see Wilkinson
     and Kitzinger 2000; Ehrenreich 2009, pp. 15–44; Rittenberg 1995, pp. 37–9;
     De Raeve 1997, pp. 249–56; Lorde 1980; Major 2002; Seager 2003.
58   Frank 1998, p. 331, emphasis in the original.
59   Frank 1993, pp. 41–5.
60   Frank 1993, p. 47.
61   Frank 1995, p. 109; Rimmon-Kenan 2002, pp. 23–4; Stacey 1997, pp. 14–15;
     Thomas-Maclean 2004, p. 1649.
62   Frank 1998, p. 338.
63   Rimmon-Kenan 2002, p. 23.
64   Jansson 1973, p. 16.

## Works Cited

Adams M. 1959, 'Truth about Cancer', *The Times*, 14 September, 11.

Aitken-Swan J. and Easson E. 1959, 'Reactions of Cancer Patients on Being Told
Their Diagnosis', *British Medical Journal*, i, 5124, 779–83.

Anonymous 1952, 'William Stewart Halsted [obituary]', *Lancet*, 259, 6718, 1100.

Anonymous 1958, 'Treatment of Breast Cancer', *Lancet*, 271, 7014, 249–50.

Anonymous 1959a, 'Surgeon Defends Shielding Patients from the Truth', *The Times*, 13 October, 7.

Anonymous 1959b, 'Breast Cancer Again', *Lancet*, 273, 7084, 1186–7.

Anonymous 1963, 'Any Questions?', *British Medical Journal*, ii, 5369, 1390.

Anonymous 2004, 'Menopausal Therapy after Breast Cancer', *Lancet*, 363, 9407, 410–11.

Anonymous 2005, 'Herceptin and Early Breast Cancer: A Moment for Caution', *Lancet*, 366, 9498, 1673.

Aronowitz R. 2001, 'Do Not Delay: Breast Cancer and Time, 1900–1970', *Milbank Quarterly*, 79, 355–86.

Baum M. and Edwards M. 1972, 'Management of Early Carcinoma of the Breast', *Lancet*, 300, 7767, 85.

Beatson G. 1896, 'On the Treatment of Inoperable Cases of Carcinoma of the Mamma: Suggestions for a New Method of Treatment, with Illustrative Cases', *Lancet*, 148, 3803, 162–5.

Broom D. 2001, 'Reading Breast Cancer: Reflections on a Dangerous Intersection', *Health*, 5, 249–68.

Cantor D. 2006, 'Cancer, Quackery and the Vernacular Meanings of Hope in 1950s America', *Journal of the History of Medicine & Allied Sciences*, 61, 324–68.

Cantor D. 2007, 'Introduction: Cancer Control and Prevention in the Twentieth Century', *Bulletin of the History of Medicine*, 81, 1–38.

Childe C. 1907, *The Control of a Scourge or How Cancer is Curable*, New York: E.P. Dutton.

Clarke M. 1998, 'Ovarian Ablation in Breast Cancer, 1896 to 1998: Milestones along Hierarchy of Evidence from Case Report to Cochrane Review', *British Medical Journal*, 317, 7167, 1246–8.

Crile G. Jr 1958, 'Treatment of Breast Cancer', *Lancet*, 271, 7023, 739.

Crile G. Jr 1972, 'The Case for Local Excision of Breast Cancer in Selected Cases', *Lancet*, 299, 7750, 549–51.

De Raeve L. 1997, 'Positive Thinking and Moral Oppression in Cancer Care', *European Journal of Cancer Care*, 6, 249–56.

Ehrenreich B. 2009, *Smile or Die: How Positive Thinking Fooled America and the World*, London: Granta.

Francis L. 1959, 'Truth about Cancer', *The Times*, 16 September, 11.

Frank A. 1991, *At the Will of the Body: Reflections on Illness*, New York: Houghton Mifflin.

Frank A. 1993, 'The Rhetoric of Self-Change: Illness Experience as Narrative', *Sociological Quarterly*, 34, 41–5.

Frank A. 1995, *The Wounded Storyteller: Body, Illness and Ethics*, Chicago: University of Chicago Press.

Frank A. 1998, 'Stories of Illness as Care of the Self: A Foucauldian Dialogue', *Health*, 2, 329–48.

Halsted W. 1907, 'The Results of Radical Operations for the Cure of Carcinoma of the Breasts', *Annals of Surgery*, 46, 1–19.

Handley W. 1905, 'Abstract of the Hunterian Lectures on the Dissemination of Mammary Carcinoma', *Lancet*, 165, 4259, 983–6.

Herndl D. 2006, 'Our Breasts, Our Selves: Identity, Community, and Ethics in Cancer Autobiographies', *Signs: A Journal of Women in Culture and Society*, 32, 221–45.

Hillman G. 1928, 'Fussing over Cancer', *Lancet*, 212, 5479, 479–80.

Holmberg L. and Anderson H. 2004, 'HABITS (Hormonal Replacement Therapy after Breast Cancer – Is It Safe?), A Randomised Comparison: Trial Stopped', *Lancet*, 363, 9407, 453–5.

Howard S. 1959, 'Truth about Cancer', *The Times*, 15 September, 11.

Jankins R. 1959, 'Truth about Cancer', *The Times*, 12 September, 7.

Jansson T. 1973, *Tales from Moominvalley*, London: Puffin.

Jordan C. 1998, 'The Development of Tamoxifen for Breast Cancer Therapy: A Tribute to the Late Arthur L. Walpole', *Breast Cancer Research and Treatment*, 11, 197–209.

Keynes G. 1928, 'Radium Treatment of Primary Carcinoma of the Breast', *Lancet*, 212, 5473, 108–10.

Keynes G. 1929, 'The Treatment of Early Breast Cancer', *Lancet*, 213, 5499, 156–7.

Keynes G. 1930, 'Mutilating Operations', *Lancet*, 215, 5549, 52–3.

Keynes G. 1937, 'The Place of Radium in the Treatment of Cancer of the Breast', *Annals of Surgery*, 106, 619–30.

Kleinman A. 1988, *The Illness Narratives. Suffering, Healing, and the Human Condition*, New York: Basic Books.

Lederer S. 2007, 'Dark Victory: Cancer and Popular Hollywood Film', *Bulletin of the History of Medicine*, 81, 94–115.

Lerner B. 2000, 'Inventing a Curable Disease: Historical Perspectives on Breast Cancer', in A. Kasper and S. Ferguson (eds) *Breast Cancer. Society Shapes an Epidemic*, New York: St Martin's Press, 25–49.

Lerner B. 2001, *The Breast Cancer Wars: Hope, Fear, and the Pursuit of a Cure in Twentieth-Century America*, New York: Oxford University Press.

Lerner B. 2006, *When Illness Goes Public: Celebrity Patients and How We Look at Medicine*, Baltimore: Johns Hopkins University Press.

Lorde A. 1980, *The Cancer Journals*, London: Sheba Feminist Publishers.

Major W. 2002, 'Audre Lorde's *The Cancer Journals*: Autopathography as Resistance', *Mosaic*, 35, 39–56.

McWhirter R. 1949, 'Treatment of Cancer of Breast by Simple Mastectomy and Roentgenotherapy', *Archives of Surgery*, 59, 830–42.

Nicolas V. 1959, 'Truth about Cancer', *The Times*, 18 September, 11.

Nissen-Meyer R. 1965, 'Castration as Part of the Primary Treatment for Operable Female Breast Cancer', *Acta Radiologica*, supp., 249.

Pantin A. 1959, 'Truth about Cancer', *The Times*, 15 September, 11.

Paterson R. and Russell M. 1959, 'Clinical Trials in Malignant Disease: Part II – Breast Cancer: Value of Irradiation of the Ovaries', *Journal of the Faculty of Radiologists*, 10, 130–3.

Patterson J. 1991, 'Cancer, Cancerphobia, and Culture: Reflections on Attitudes in the United States and Great Britain', *Twentieth Century British History*, 2, 137–49.

Platt R. 1959, 'Truth about Cancer', *The Times*, 17 September, 11.

Rabinovitch D. 2007, *Take Off Your Party Dress: When Life's Too Busy for Breast Cancer*, London: Pocket Books.

Rimmon-Kenan S. 2002, 'The Story of "I": Illness and Narrative Identity', *Narrative*, 10, 9–27.

Rittenberg C. 1995, 'Positive Thinking: An Unfair Burden for Cancer Patients?', *Supportive Care in Cancer*, 3, 37–9.

Robinson J. 1986, 'Treatment of Breast Cancer through the Ages', *American Journal of Surgery*, 151, 317–33.

Rouse A. 1959, 'Truth about Cancer', *The Times*, 15 September, 11.

Rowe R. 2003, 'Bile Beans for Inner Health', *International Journal of Pharmaceutical Medicine*, 17, 137–40.

Seager J. 2003, 'Rachel Carson Died of Breast Cancer', *Signs: A Journal of Women in Culture and Society*, 28, 945–72.

Segal J. 2007, 'Breast Cancer Narratives as Public Rhetoric: Genre Itself and the Maintenance of Ignorance', *Linguistics and the Human Sciences*, 3, 3–23.

Sinclair E. 1959, 'Truth about Cancer', *The Times*, 18 September, 11.

Stacey J. 1997, *Teratologies: A Cultural Study of Cancer*, London: Routledge.

Standard S. and Nathan H. 1955, 'Preface', in S. Standard and H. Nathan (eds) *Should the Patient Know the Truth?*, New York: Springer, 9–10.

Thomas-Maclean R. 2004, 'Understanding Breast Cancer Stories via Frank's Narrative Types', *Social Science and Medicine*, 58, 1647–57.

Toon E. 2007, '"Cancer as the General Population Knows It": Knowledge, Fear, and Lay Education in 1950s Britain', *Bulletin of the History of Medicine*, 81, 116–38.

Wilkinson S. and Kitzinger C. 2000, 'Thinking Differently about Thinking Positive: A Discursive Approach to Cancer Patients' Talk', *Social Science and Medicine*, 50, 797–811.

# 3
# Running Out of Options: Surgery, Hope and Progress in the Management of Lung Cancer, 1950s to 1990s

*Carsten Timmermann*

When Frank Craig died in March 1994, almost exactly four years after a third of his right lung had been removed by a surgeon at St Anthony's Hospital in Cheam, south of London, his death was sadly representative of the fate of many lung cancer patients in Britain.[1] About half of the patients operated on for lung cancer die within five years after the operation. This figure is lower than it used to be, but not because of new, innovative treatments. The percentage of long-term survivors after surgery is higher now than in previous decades mostly due to the introduction, since the 1970s, of more rigorous methods of distinguishing patients for whom surgery may bring long-term survival from those to whom an operation would be of no use. In recent years, less than a quarter of patients diagnosed with lung cancer have been undergoing lung resections, the removal of parts or all of one lung. Around the time of Frank Craig's illness, about 10 per cent survived their lung cancer diagnosis for five years or longer – usually following surgery.[2]

Surgery has been the main stay of lung cancer therapy ever since bronchial carcinoma turned from a rare disease into a major killer in the mid-twentieth century (no other cancer has a higher incidence) and both procedures and survival rates have changed surprisingly little since lung resection became routine in the late 1940s.[3] Some thoracic surgeons in Britain continued until the 1980s with treatment policies formulated in the 1950s.[4] There was a moment in the mid to late 1960s when chest surgeons appeared to feel that, as far as lung cancer surgery was concerned, history had ended: they no longer expected significant innovations in the surgical treatment of carcinoma of the bronchus (and neither in radiotherapy, for that matter) and were doubtful about attempts by younger colleagues to introduce ever more sophisticated methods of diagnosis designed to identify those cancers that were operable.[5] If a patient was fit enough for an operation, the operation should be performed without delay. If there was any hope for innovations, then this lay not in surgery but in fundamental research on the biology of cancer. Conventional lung cancer therapy, organised around the resection of the affected parts of the lung and adjacent tissues was as good as it could possibly get, 'standardized, effective and efficient'.[6]

Pessimism about the possibility of continuing progress (or at least the open confession to such pessimism) is unusual in modern medicine and, especially in cancer treatment we have come to expect a commitment to research and permanent improvement. Routine, it seems, is not good enough. We expect practitioners and patients to 'battle' cancer, to never give up in the quest for new solutions that delay death (the use of battlefield vocabulary and its implications have been discussed by the journalist John Diamond in his account of his experiences while being treated for cancer of the tongue and throat).[7] Doctors, patients and carers have different ways of dealing with lack of progress, which they may or may not perceive as defeat in a battle. I will draw in this chapter on the story of Frank Craig's lung cancer as told by his wife, the religious writer and broadcaster Mary Craig, using her account of her husband's illness and death as a case study to reflect on the treatment of lung cancer in the post-war period. I am interested in exploring a patient's perspective on a pathway that has become so standardised that it can be depicted as a fairly simple flow chart (Figure 3.1). To give my case study context, I will survey briefly the development of (surgical) treatment pathways for non-small cell lung cancer up to the 1990s and discuss the origins of the lack of therapeutic enthusiasm expressed by thoracic surgeons in the 1960s, and its consequences for patients.[8]

History, done well, is as much about meanings as events. Diseases, as historians of medicine have long argued, change their meanings over time and place, in line with the cultural values of the societies and social groups that they affect, and as an effect of the technologies (in the widest possible sense) that are appropriated and invented to deal with them. In this chapter I want to explore what it does to the meaning of a disease, to social interpretations, the expectations of practitioners and the hopes of patients, if this disease acquires the label of hopelessness, if little changes in the way it is treated, during a period when much appears to be changing in medicine otherwise. The ways in which the developed world has dealt with other diseases, including some malignant diseases, have changed significantly since the 1940s. For example, childhood cancers or leukaemias and lymphomas were declared curable by way of chemotherapy in the 1970s, and this changed their meanings fundamentally.[9]

How does one write the history of routine procedures, the unexciting stuff where not much changes and where professionals do not make their names with pioneering innovations? Professionals reflect on these issues, at least occasionally, and thereby make it easier for historians to reconstruct their side of the story. It is more difficult to get to patients' perspectives. We could rely on the brief case descriptions that surgeons include with their publications, but they lack fine grain, are told with hindsight, and more importantly, interpret the patients' stories through the lens of the surgeons. The ideal sources for us to learn what it meant for patients to be diagnosed with and treated for lung cancer, are diaries, memoirs or ethnographic studies.[10] As my reading of Craig's account of her husband's illness illustrates, events and developments that are routine for professionals, are meaningful and unique for patients and carers. While a large number of cancer memoirs have been published in recent years (we have discussed this in the introduction to this volume), unfortunately, next to none deal with the experience of lung cancer. This may well be due to the fact that lung cancer affects predominantly elderly men, often working class, who rarely write diaries or blogs.[11] Frank Craig's story of lung cancer in the early 1990s, one of the few memoirs that deal with this particular cancer, allows me to illustrate the patients' side of a development that was well underway in the late 1960s. Mary Craig's book, moreover, based on the diary she kept at the time of her husband's illness, is particularly insightful, perhaps more so than other books in the genre.

## Lung cancer in late twentieth-century Britain

Frank Craig was 63 when he was first diagnosed and 67 when he died – after what was a fairly normal lung cancer pathway, at a normal age for a lung cancer patient.[12] Mary Craig does not tell us if her husband smoked, but the great majority of lung cancer patients have been smokers or ex-smokers. The causal link between lung cancer and cigarette smoking has been more or less generally accepted since the work of Doll and Bradford Hill in Britain, and Wynder and Graham in the United States in the 1950s, and the publication in the early 1960s of reports by the Royal College of Physicians and the US Surgeon General.[13]

The Craigs first suspected that something was wrong when Frank coughed up blood in July 1990. Mary Craig does not tell us anything about the investigations that followed until January 1991, when they were told that Frank may have lung cancer: 'It's only a possibility – the merest hint of a shadow on an X-ray picture – and we shan't know anything more till next week after he's had a bronchoscopy'.[14] A biopsy revealed cancerous cells and Frank was referred to Tom Treasure, 'a highly reputed surgeon in Wimpole Street', who was '90 percent sure Frank had lung cancer, and advised the removal of a section of the right lung'.[15] Treasure described Frank's chances of survival as 'Better than evens in your case', which shocked the Craigs, as the first surgeon they consulted had talked about 'total cure'.[16] Treasure has reflected on this apparent contradiction between his prognosis and that of the first surgeon in an email to the author, suspecting that it was an expression of a 'conspiracy of optimism in cancer talk between doctors and patients'.[17] It may also have been, he suggests, 'part of a strategy, not necessarily deliberate and sometime forced on us by circumstances of what John Diamond calls 'the principle of gradual disclosure'.[18]

Frank Craig's operation took place on 4 March 1990 and Treasure phoned the Craigs a week later with the pathology results. He assured them, to their relief, that

> he had cut out all the cancerous tissue, there were no secondaries and apparently no cancer in the bloodstream. A few lymph glands which had been infected had been taken away. He could only hope that they'd all been caught, that there were no rogue ones that might have escaped into the bloodstream.[19]

To Mary Craig, Treasure's statement sounded like an expression of confidence. To the surgeon, reflecting in retrospect, the statement shifted the cancer from a hoped for stage I (no lymph nodes affected; greater chance of cure) to a stage II or even IIIa, lowering the statistical chance of a positive outcome significantly.[20] To Mary Craig, the reassuring tone in his voice was more important than the implications of what he said (which possibly were not spelled out to her explicitly). A friend, however, told her that she should not take too much for granted and make the best of the year or two she and Frank may have left. 'I was a bit sore at the implication that Frank was henceforth living on borrowed time. How rapid the sense of danger departs once the all-clear sounds!'[21] Mary Craig hoped that the story of her husband's lung cancer was over: 'I *do*, absolutely unequivocally, hope for a cure. I want out of all this'.[22] A diagram depicting a normal treatment pathway for non-small cell lung cancer in a book on lung cancer, written by the British lung cancer specialist Chris Williams for patients and their relatives, appears to support such hopes.[23] The box labelled 'Operation' looks a bit like a door; the door that allowed patients to escape from their lung cancer: a way out (Figure 3.1). Indeed, the removal of part of the lung or the whole lung is the therapeutic intervention that for the last five decades has offered the best chances for a permanent remission for non-small cell lung cancer patients.

*Figure 3.1*   Schematic depiction of a 'normal' treatment pathway for non-small cell lung cancer. Reproduced from Chris Williams, *Lung Cancer: The Facts*, 2nd edition, Oxford: Oxford University Press, 1992, p. 51. Reproduced with permission.

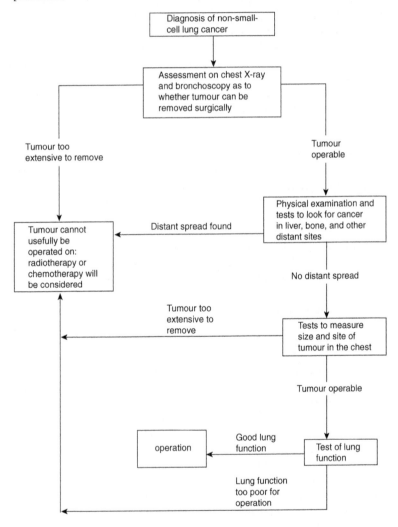

Only for a minority of patients, however, did the story of their cancer end with surgery. With good reasons we do not find references to 'rates of cure' in scholarly publications on cancer therapy, just 'survival rates'.[24] Survival rates after surgery changed very little from the 1950s, when lung resections became routine, to the 1980s. In a study published in 1983 by J. R. Belcher, comparing the long-term results of several prominent British surgeons, survival rates varied between 25.5 and 26.8 per cent after five years, and between 13.6 and 17.8 per cent after ten years, with surprisingly little change since the 1950s.[25] Belcher himself, thoracic surgeon to the London Chest and Middlesex Hospital, recorded a five-year survival rate of 28 per cent for the period from 1950 to 1955 and of 27 per cent for 1970 to 1975, while the ten-year survival rate remained unchanged at 15 per cent. The rates were similar for conservative surgeons and for their more aggressive colleagues and it did not make any difference for survival if lobectomies (resection of parts of a lung) or pneumonectomies (resection of a whole lung) were performed.[26] To Belcher the 'disease process itself' seemed to be the dominant factor that determined survival rates.[27] While the incidence of lung cancer continued to rise and chest surgeons continued to play a central part in the treatment of lung cancer, there was increasingly less hope among them that technical progress was possible, and this as early as the mid-1960s. In fact, as we will see in the next section, there was a distinct sense of disillusionment developing among the professionals.

## 'Almost unmitigated gloom': The 1960s and the end of enthusiasm

'The contribution that surgeons can make to this subject, in so far as the solution of the problem is concerned, has already been made', Mr N. R. Barrett (Surgeon to St Thomas's Hospital and Brompton Hospital, and Consultant Surgeon, Royal Navy, King Edward VII Hospital, Midhurst and Ministry of Pensions) suggested in the introductory address to a symposium on the diagnosis and treatment of carcinoma of the bronchus in 1966, 'and I do not believe there is much more any surgeon can add with advantage'.[28] After evaluating the results of 40 years of surgery for bronchial carcinoma, thoracic surgeons now were at a stage where 'the *raison d'être* should be discussed – why we are doing surgery at all'.[29] He viewed the idea of a 'cancer operation', the radical removal of a tumour and its lymphatic field in one block, as suspect. Just because it worked for breast and rectal cancer, this did not mean that the concept was applicable to lung cancer. Neither did he think that diagnosis would make a difference – 'Is there any evidence that improved methods of diagnosis would help the results of surgery? I think there is none' – or that early detection was the answer: 'I would suggest to you that the conception on which surgeons have worked – namely that if they are sent the patients early their results will be good – is nonsense'.[30] Tinkering with surgical procedures was pointless: 'I think the technique of surgery in relation to immediate survival after operation for carcinoma of the bronchus is standardized, effective and efficient'.[31] Barrett did not stop with surgery: 'Radiotherapy in the treatment of malignant disease has played its part' and '[c]onventional pathology also has little more to offer'.[32] New solutions, Barrett argued, could only be expected from laboratory scientists. Surgeons had to get up to speed with modern cancer research and biology rather than hope that improvements of technique were going to make any difference. Surgeons, however, Barrett suggested, while not likely to do any further promising research themselves, still played an important role as clinicians who helped laboratory researchers with careful observations of tumour development and its clinical manifestations, 'in recording the exceptions rather than the rules'.[33]

Barrett was not alone with his thoughts about the end of history in thoracic surgery. P. R. Allison of Oxford confessed to similar doubts in an address dealing with 'The Future of Thoracic Surgery', delivered to the Society of Thoracic Surgeons of Great Britain and Ireland in October 1965. 'Carcinoma of the lung is treated by surgery now as well as it ever will be treated by surgery', he argued. 'This is a technical exercise, but the ultimate solution to carcinoma of the lung must surely be a biological one and not a surgical one'.[34] Neither was this attitude unique to Britain. In the discussion following Barrett's opening speech at the Midhurst symposium, J. S. Chapman of the South Western Medical School, University of Texas, agreed with the speaker and added that '[t]he point of view Mr Barrett has expressed this morning is almost identical with that of many other thoughtful surgeons in the United States'.[35] Another participant, Jack Belcher, characterised the predominant mood during the symposium as 'almost unmitigated gloom'.[36]

Surgeons had developed the first promising operations for this once-rare disease in the 1930s. Since then they had 'owned' lung cancer, controlled the procedures and overseen the progress of patients.[37] Opening the thorax confronted surgeons with a formidable challenge and, as Pack and Ariel put it in their celebratory account of *A Half Century of Effort to Control Cancer*, '[t]he chest was one of the last anatomic regions to be surgically invaded'.[38] Overcoming these problems, with great difficulty and with the help of modern technology, was a source of pride and professional power for surgeons. After experiments on animals, surgeons performed the first pneumonectomies and lobectomies on human patients in the early 1930s. The credit for the first successful pneumonectomy for lung cancer went to the American surgeon Evarts Graham who performed the operation in 1933. The patient, famously, outlived his surgeon, a lifelong heavy smoker, who died himself from lung cancer in 1957.[39]

In the late 1940s, when lung cancer incidence increased dramatically, the operations became routine. The most important factor in making this possible was improvement in anaesthesia practices. The most significant technical innovations were the endotracheal tube and the blown rubber bag which enabled selective ventilation of a single lung. Curare and other drugs that paralysed the autonomic nervous system further simplified operations. Tuberculosis was now treated with antibiotics and became less common, and lung cancer surgery became the bread and butter operation for thoracic surgeons. Surgeons also controlled the diagnosis of lung cancer. When a patient was referred to them with a suspicious shadow in the chest X-ray or a persistent cough, they performed the obligatory bronchoscopy and secured the tissue samples that were sent to the pathologists. When depression about the stagnation in lung cancer surgery overcame them in the 1960s, thoracic surgeons increasingly saw heart surgery as the new challenge, as the field where fame and power could be gained. Around the same time, chest physicians assumed a somewhat more central role in the diagnostic pathway, which they cemented with the introduction of the flexible fibre-optic bronchoscope in the early 1970s, operated by physicians more frequently than surgeons.[40] Medical oncologists also developed an interest in lung cancer and adapted chemotherapy regimes for this disease, but there was no major breakthrough, and the mainstay of the curative (as opposed to palliative) treatment of non-small cell lung cancer remained surgery.[41]

Not all thoracic surgeons agreed with the pessimistic outlook of Barrett and Allison. Dissenting voices could be heard even in the deep gloom of the Midhurst meeting. Jack Belcher, for example, then surgeon to the London Chest Hospital, Thoracic Surgeon to the Middlesex Hospital and Consultant Thoracic Surgeon to the North West Metropolitan Regional Hospital Board, confessed that after one and a half days of listening to gloom he 'felt increasingly proud to be a surgeon since we at least can do something for this disease'.[42] Belcher, it seems, and a small number of colleagues, maintained their enthusiasm until they retired in the 1980s, by which time they were veterans of their trade.[43]

There was in fact some progress that could be recorded since the 1950s. The results that Belcher and his colleagues reported were significantly better, for example, than what the British Empire Cancer Campaign (BECC) survey of cancer in London revealed in 1952, when out of 1024 lung cancer patients 178 were operated on and only one patient survived the operation for five years or more.[44] Most London surgeons then had obviously not reached the high standards of Belcher and his colleagues. Out of the same patient population, 239 were treated by radiotherapy, and only four of these survived five years or longer. Six hundred and seven were treated neither by surgery nor radiotherapy. These figures made 25 per cent survival look very desirable. The difference, it seems, was the routine achieved by Belcher and his colleagues and their specialist focus on chest surgery. The challenge now was administrative and political rather than technical. How could good routine treatment be made available to as many patients as possible as quickly as possible?

The opposite of routine is probably the clinical experiment. Much cancer treatment has been experimental throughout the twentieth century.[45] Radiotherapy in the interwar period, for example, then a new and exciting modality, led to the establishment of new cancer centres. At Manchester's Christie Hospital and Holt Radium Institute, for instance, under the leadership of the Scotsman Ralston Paterson and his wife Edith, experimental approaches to treatment went hand-in-hand with the development of new statistical methods and experiments with innovative forms of teamwork in therapy. This turned the Christie into a pioneering institution, both organisationally and in terms of the methods created and the knowledge produced. In the post-war period, increasingly, chemotherapy was the form of treatment that was expected to bring new breakthroughs for the cure of cancer. Chemotherapy had been successful in overcoming infectious disease, and the expectations created by the successful streptomycin trials in the treatment of tuberculosis shaped the plans of the British Medical Research Council (MRC) for cancer research. But randomised clinical trials, which worked well for testing new drugs especially in previously untreatable conditions such as leukaemia, did not work so well where a reasonably reliable treatment modality was available, and particularly where this was surgery.[46]

Using any modality other than surgery as a first-line therapy was ethically questionable as soon as surgery had become routine, which made it almost impossible to organise clinical trials comparing different therapies, unless they were adjuvant to surgery or targeted patients who some frustrated trial organisers termed 'surgical rejects'. In a rare trial in the mid-1950s, involving the other established locally acting treatment modality, radiotherapy (rather than systemic, like chemotherapy), Gwen Hilton, radiotherapist at University College Hospital London, treated patients suffering from operable lung cancer and, in fact, obtained results with ordinary deep X-rays (not the latest machines) that were very similar to those that the surgeons achieved.[47] Ten years later, during the gloom of the Midhurst symposium, her co-author Joseph Smart, physician at the London Chest, the Brompton and the Connaught Hospitals reminisced:

> The problem ... was to get the cases. When I suggested the investigation to Dr Hilton she said she would like to do it but she never got these cases. This was very understandable as any surgeon seeing an operable carcinoma naturally has to take it out.[48]

The MRC treatment trials for lung cancer in the 1960s and 1970s were disappointing, for surgery as well as radiotherapy and (adjuvant) chemotherapy.[49] Indeed, they added to the growing frustration of the surgeons, particularly because in the first of these trials radiotherapy seemed to produce marginally better outcomes than surgery. There were a number of reasons for this, the most important being that the trial involved patients suffering from small cell carcinoma, a particularly aggressive variant that metastasised quickly and was usually considered inoperable. However, to superficial readers the trial results may have seemed to apply to lung cancer in general – especially if they only read the discussion on the correspondence pages of the *Lancet* following the publication of the first trial report, where one author of the study wrote about bronchial carcinoma as a disease where the focus should be on prevention rather than cure.[50] It seems that the publicity generated by a clinical trial can do as much damage as good, and in the interest of treating larger numbers of patients successfully, it may sometimes be better to put resources into securing broad access to routine techniques, ways of choosing the patients likely to benefit.

Today, four decades later, according to Tom Treasure, typically about half of the patients who undergo surgery survive the operation for five years or longer, compared to the 25 per cent reported by Belcher.[51] But this improvement has been mostly due to a more effective selection process following the development in the 1970s and fairly broad international adoption in the 1980s of a standardised system for staging lung cancer, based on the so-called TNM (Tumour-Nodes-Metastases) system embraced by the International Union against Cancer (UICC).[52] The wider availability of new diagnostic technologies such as CT and PET scanners improved assessment, and internationally agreed staging systems provided them with clear, standardised criteria, which allowed the comparison of results. Much effort has gone into identifying patients who were likely to benefit from surgery and ruling out those who were not. By the late 1990s, about half of the patients were ruled out after the first set of examinations, because they were too weak or had already developed metastases. After further tests, which usually included a bronchoscopy and often a scan, about 25 per cent of the patients presenting with a non-small cell lung cancer underwent surgery.[53] If about half of these survived for five years or longer, this equalled a good 10 per cent of all patients with a confirmed diagnosis of lung cancer. Unfortunately we cannot compare these figures with Belcher's, as we do not know what proportion of patients diagnosed with lung cancer went on to surgery. However, during the registration years 1971–1973 the crude survival rate for all male lung cancer patients in Britain was 19 per cent after one year and only 6 per cent after five years, and the prospects were slightly worse for women.[54]

What does lack of progress in the therapy of a particular malignant disease mean for individual patients? How do the expectations shaped by interactions with surgeons and physicians, who know the figures, translate into patient experience? In the next section we return to the story of Frank Craig, who found himself among the quarter of all patients fit to undergo surgery but who, as it turned out, was not among the lucky 10 per cent who survived for five or more years. Treating the Craigs' experience as the historical equivalent of what sociologists call 'a qualitative single case' study will provide us with insights (some very personal) into a part of the lung cancer pathway on which surgeons rarely comment.[55] What happens if the cancer returns is difficult, maybe impossible to standardise. Looking at individual, historical patient experiences like those reflected on by Mary Craig may be a good way towards understanding how patients and their relatives maintain hope and defend normality in situations that frequently confront them with existential challenges.[56] What happens if the 'exit door' in the flowchart does not lead away from cancer?

## Running out of options

For Frank Craig the story of his lung cancer was not over after the operation. Surgery had not cured him. The cancer reappeared in 1993, when a new chest physician had just told him that after almost five years he 'was likely to be soon off the danger list as far as cancer was concerned'.[57] The first signs were back pain in summer 1993 (which did not overly concern the Craigs, as they did not associate it with his lung cancer), breathlessness and exhaustion during a cruise in February 1994, dizziness in April, increasingly bad problems with passing water and pains in the prostate. In July, a small growth in Frank's neck was detected, which proved malignant. How much time did he have left? 'Dr L. could not be sure, of course, but thought six, maybe nine months. From the way he didn't quite look at us as he spoke the words, I knew we had nothing to hope for'.[58] Dr L. told them that nothing could be done: 'Chemotherapy was largely a waste of time in the case of lung cancers and would involve a lot of extra angst for nothing'.[59] While they knew that they had nothing to hope for, however, they continued to hope. And not only Frank and Mary Craig, the doctors and nurses who treated Frank also seemed to operate with ambiguous expectations and notions of hope.[60] The boundary between palliation and curative treatment started to blur, and this blurring formed a precondition for the continuation of treatment; it maintained hope and the motivation to go on. The clean distinction between palliation and curative intention suggested by protocols and treatment schemes may in practice be difficult to maintain. Predominantly palliative treatments, although the main emphasis lies on the prevention of pain and suffering, often also prolong patients' lives. Only in textbooks is the boundary clear.

Further tests failed to reveal where Frank Craig's cancer was hiding. 'We have just returned from the hospital where we were stunned to learn that the latest X-rays show no sign of cancer anywhere. There must be some somewhere – the biopsy proved that – but it now seems unlikely that Frank is riddled with it. It's all very perplexing'.[61] The doctor put things into perspective by suggesting this meant they might have a year rather than months. Madeleine, the Macmillan nurse, made her first visit and reinforced this view. Craig quotes her as saying 'Secondaries are secondaries, ... and they are incurable'.[62] The prostate problems got worse and a prostatectomy was performed on 9 August. Friends put the Craigs in touch with Raoul Coombes, Professor of Oncology at Charing Cross Hospital: 'We have an appointment to see Raoul – positively our last hope – on Friday'.[63]

Tests at Charing Cross revealed the only lump to be found in a gland near the one removed from Frank's neck in July, and that was 'small and thankfully dormant'.[64] On 19 September another bronchoscopy was performed. 'The bronchoscopy ... was a nightmare experience, and Frank said afterwards that if he'd known how awful it would be he'd have refused point-blank to have it. We're getting sick to death of hospitals, yet there's no escaping them'.[65] On 21 September further neurological tests followed, and the results of the bronchoscopy. The airways were narrowed and inflamed, but not completely blocked. Dr Lim, a doctor who they had not seen before, 'didn't pull any punches. The lung cancer had spread into the system, he said, and nothing could halt its advance. Chemotherapy would be a waste of time'.[66] Steroids might help in the short run. 'Raoul joined us and confirmed that they had run out of options'.[67]

The steroids had little effect, except that Frank was slightly less breathless. He continued to suffer from loss of balance, lack of energy, back ache and prostate trouble. 'Madeleine, our Macmillan nurse, came in today. We didn't ask her to give us any hope, because we know there isn't any'.[68] A bone scan on 26 October was the next attempt to find the cancer. 'Frank can scarcely stand unaided now and really can't be left alone for a moment. Raoul, when we saw him on Monday, was obviously shocked by the rapid deterioration in his condition. ... *something* clearly is far from dormant'.[69] But, really, despite the repeated assurances that there was none, nobody seemed to give up hope, neither the doctors, nor the Craigs. The doctors continued to search for the tumours in the hope that then at least they could palliate, and the Craigs hoped that Frank at least would be able to walk again, would suffer less pain, and would be less helpless. The hope for survival, it seems, also survived.

Most diagnostic tests were negative, until finally a lumbar puncture revealed an active tumour in the spinal fluid. 'Frank rang me from hospital, actually quite relieved that something positive had at last come to light. Raoul, he said, was cautiously optimistic'.[70] After having been told repeatedly that 'chemotherapy would be a waste of time', Frank now received several courses. But the tumour did not respond and the chemotherapy was discontinued. On 24 November, then, two cancerous nodules in the spinal column were identified and the doctors decided to try radiotherapy, starting on 2 December. Two weeks later, the hospital staff told Frank that 'the radiotherapy obviously wasn't working, and they had nothing more to offer'.[71] When returning home he broke down and cried (the first time his wife saw him crying). 'I don't know whether it was this sudden removal of hope, or the fact that he had to come home and break the news to me, that had caused him to break down'.[72]

> Well, it's reappraisal time again. We came to terms with the original state of affairs when it hit us some months ago, and now we must do so again. Why isn't it possible to do so once and for all when trouble first strikes? Why do we always have to keep picking ourselves up and starting again from the beginning? I suppose it is because hope will insist on springing eternal and then we drop our guard. Once again we had been clutching at straws – we'd sent out a couple of hundred of Christmas letters to friends – (well, two hundred copies of one letter), saying how hopeful we were about the radiotherapy. (Actually, in

drafting the letter, I had typed that we had 'great hopes' of it, but Frank had made me change that to 'some small hope'.) But that was then and now is now. The markers are being called in.[73]

When he went to bed that evening, Frank Craig predicted that in two weeks he would be bed-ridden. On 14 November he wrote a note to Raoul, suggesting that he might not be able to undertake the journey to Hammersmith for much longer. The note ended: 'It looks as though we have lost the battle, Raoul, but I am grateful to you for trying so hard to win it'.[74]

Frank's illness progressed, apparently unstoppable. On 19 December he went to Charing Cross Hospital one more time, 'but mechanically now, with nothing left to hope for'.[75] Madeleine, the Macmillan nurse, phoned Mary after having spoken to Raoul: 'She says the next thing will be palliative care. It's all that's left – making him as comfortable as possible while waiting for the end'.[76] But was it not palliative care from the moment the secondary tumours were diagnosed? In the Craigs' life world, evidently, palliative care did not have clearly defined boundaries. The treatment had meanings for the Craigs, and these meanings over-lapped and sometimes contradicted one another. The treatment became more palliative every time hope vanished, and more curative every time hope returned. 'Madeleine won't hazard a guess about his life expectancy. Even now, she says, the cancer could go into remission, in which case it would be neither better nor worse, just static'.[77] A duty nurse at Charing Cross told Frank that sometimes one cannot see any results till a week or two after the end of the radiotherapy. 'She meant well, I'm sure, and the fragile hope she offered seems to have cheered him up, so I'll have to play along with it. But I dread his next appointment in January when he will inevitably discover that nothing has changed'.[78] On 22 December Raoul met Frank with the words, 'It seems you have us beat, Frank', but asked to see him again in January, in case the radiotherapy took belated effect.[79]

The radiotherapy did not take belated effect, and the illness progressed further. Frank Craig found it harder and harder to leave his bed and by the second half of January was sleeping most of the time. The Craigs were provided with a growing number of 'gadgets', as Mary Craig calls them, walking frames, bath contraptions and syringe drivers, which helped with Frank's care. They also needed (and were provided with) increasingly more comprehensive nursing cover, first just on weekdays, then for the weekends and later also at night. It was now clear to everybody, it seems, that the decline was final, progress inevitable. 'Every event seems to be another "last time". Another final curtain ringing down'.[80] Frank talked about his funeral and started to read the obituaries in the *Times* every day, and he admitted to Mary that he was discouraged to note that most people were older than his 67 when they died. 'I suppose it's natural for us to feel resentful when death's shadow begins to fall – we're never really ready to let go of life'.[81] Frank grew more and more passive. 'Hanging on. Waiting. It's what our lives largely consist of now'.[82] He was also getting more erratic, drifting in and out of lucidity. The prostate caused new problems and pain, and a catheter had to be inserted with great difficulties. 'Poor Frank is being subjected to one physical indignity after another. In the kindest possible way, of course, but it's a chastening thought that most of us come to such ritual humiliation at the end. Fortunately I don't think he knows much about it'.[83] On 6 March Frank Craig stopped breathing, two days after the fourth anniversary of the lung cancer operation, 'which had at the time seemed so successful'.[84]

## Some reflections on progress and routine, cure, survival and palliation

Along with nine out of ten of all lung cancer patients in Britain, and about half of those who are found fit for lung surgery, Frank Craig did not survive the first five years after his diagnosis. His experiences illustrate how fuzzy the boundaries can be between palliative and curative treatment for patients confronted with this illness. It is evident from Mary Craig's account, however, that to the professionals treating Frank Craig, surgery (and only surgery) carried connotations of cure, and the other modalities, radiotherapy and chemotherapy were used when the treatment became predominantly palliative.

Hope for survival was associated with surgery, 'cutting out the cancer', and hope for alleviating suffering on the path to inevitable death with these other modalities. These associations have been in place without much change since the 1950s, despite the success of the hospice movement and the vast improvements in palliative care. The lack of breakthrough innovations in the treatment of lung cancer dispirited chest surgeons since the mid-1960s. The improvement of five-year survival rates from 6 per cent in the early 1970s to 10–12 per cent in the late 1990s, it seems, did not impress anybody sufficiently. Doctors often experience the deaths of patients in their care as personal failures.[85] Lung cancer was increasingly perceived as a 'Cinderella Cancer', neglected by researchers and policy makers due to its link with smoking and the corresponding stigma attached to the disease.[86]

It is quite evident, from the accounts of practitioners concerned with the development of their field and patients undergoing treatment, that patients and doctors ascribed different meanings to progress and stagnation. The notion of stagnation in surgical technique is obviously worse for surgeons than for patients, for whom routine and the safety that comes with it are distinct advantages. Research, conducting a trial and publishing the results, for surgeons and other progress-minded practitioners means overcoming stagnation, pushing the limits. Progress is personal, and can be measured in terms of publications, techniques or technologies associated with one's name, career progression, improvement of results over time, and this is what the 'battle against cancer' meant (and probably still means) for surgeons and physicians.[87] Progress creates career pathways. If there was no scope left for publishable improvements in lung cancer treatment, there was not much scope for self-promotion either. Since the 1960s, chest surgeons were pushing for limits mostly in heart surgery, where results were more rewarding, and settled for the apparently inevitable in lung surgery.

For patients, progress is also personal, along a different sort of pathway, where the long-term improvement of surgical technique (or other modalities) is not an issue. Patients do not look at pathways from the same perspective as the professionals treating them; a patient experiences one unique, very individual pathway. Experiences are hard to standardise. Some long-term survivors suffer from depression, others opt out of curative treatment, and neither of these fit into the standard scheme with its focus on survival.[88] Trials and research, however, have also become central to many patient pathways.[89] Only fairly recently, though, has it become desirable for patients to be subjects in clinical trials, and the recent history of cancer research may have played its role in this change. Patients volunteer for trials because they assume that the latest treatment, about which least is known, is also the best, or because they see participating in these trials as their last hope.[90] However, only trials, or better still, years of experience can reveal if this is really the case. This is a dilemma. As we have seen in Frank Craig's case, however, there is always progress in individual patient pathways, decisions to be taken, choices to be made. Progress, in fact, can also mean running out of options.

## Acknowledgements

This work was funded by the Wellcome Trust as part of the Programme Grant 'Constructing Cancers, 1945–2000'. I am grateful to Tom Treasure for his help and his historically and surgically well-informed comments on an earlier version of this essay. I would also like to thank my colleagues Emm Barnes, John Pickstone and Elizabeth Toon for their comments, as well as the participants in the Patients and Pathways workshop.

## Notes

1  Craig 1997.
2  Williams 1998; Martini 1993; Treasure, personal communication.
3  Belcher 1983; Pack and Ariel 1955.
4  Belcher 1983; Belcher 1956; Abbey Smith 1957.
5  Allison and Temple 1966.
6  Barrett 1966, p. 1.
7  Diamond 1998, p. 72.
8  Non-small cell lung cancer is a classificatory category based on treatment. Since the 1960s there was increasingly broad consensus that patients suffering from small or oat cell carcinoma did not benefit from surgery, but that tumours of this cell type, which tended to metastasise quickly, responded reasonably well to chemotherapy, albeit temporary. Non-small cell lung cancers comprise those cell types that are still treated by surgery: squamous cell, anaplastic and adenocarcinomas.
9  Krueger 2008; Barnes 2005 and 2007.
10  For an insightful ethnographic study on patients undergoing chemotherapy for small cell lung cancer in a clinic in the Netherlands and the professionals treating them, see The 2002.
11  In a survey undertaken by the BECC in London in 1952, the largest cohort of lung cancer patients belonged to social class III (skilled occupations). By 1970 this had changed and the largest group of lung cancer patients were from social class V (unskilled labourers). See Harnett 1952, p. 112; Doyal *et al.* 1983, p. 14; Williams 1992.
12  Cf Williams 1992, p. 3. The age group between 60 and 69 was also the largest among the lung cancer patients operated on, for example by the thoracic surgeon Michael Bates, between 1972 and 1983: Bates 1984. I should also mention, however, that Craig (like Diamond) had access to private health care, his pathway, thus, was not that of a normal NHS patient.
13  Lock *et al.* 1998.
14  Craig 1997, p. 7.
15  *Ibid*, p. 9.
16  *Ibid*.
17  Treasure, personal communication.
18  *Ibid*; Diamond 1998, p. 114.
19  Craig 1997, p. 12.
20  Treasure, personal communication.
21  Craig 1997, p. 12.
22  *Ibid*.
23  Williams 1992.
24  I don't have the space here to talk about the experience of long-term survival after surgery – this may be material for another paper.

25 Belcher in 1983 commented that, 'As bronchial carcinoma is so common and surgery offers almost the only hope of prolonged survival, it is surprising that so few large series have been reported'. Belcher 1983, p. 430.

26 It did make a difference for operative mortality, though. More patients died after pneumonectomies.

27 Belcher 1983.

28 Barrett 1966, p. 1.

29 *Ibid.*

30 *Ibid*, pp. 1–2.

31 *Ibid*, p. 2.

32 *Ibid.*

33 *Ibid*, p. 4.

34 Allison and Temple 1966, p. 100.

35 *Ibid.*

36 Belcher 1966, p. 100.

37 Barrett, for instance, began his address with the following words: 'It is my function to introduce the subject of carcinoma of the bronchus in a general way, and I am well chosen for this because I am a surgeon'. Barett 1966, p. 1. For the history of thoracic surgery, see also Mountain 2000; Hurt 1996; Abbey Smith 1982; Meade 1961.

38 Pack and Ariel 1955.

39 Mueller 2002.

40 Ikeda 1970; Mitchell *et al.* 1980.

41 Thatcher and Spiro 1994.

42 Belcher 1966.

43 One of these colleagues was Roger Abbey Smith, who has written a useful overview article on the history of lung surgery in Britain: Abbey Smith 1982. Both were members of Charlie's Club, a small society of thoracic surgeons whose minute books can be viewed in the Archives of the Royal College of Surgeons in London. On the history of another of these clubs, see Milstein 1991.

44 Harnett 1952, p. 119.

45 Pickstone 2007.

46 Timmermann 2007; Valier and Timmermann 2008.

47 Smart and Hilton 1956. This seems to support Belcher's fatalistic notion that it was the 'the disease process itself' that decided the outcome of therapy, more than the modality chosen.

48 Contribution by Smart to the discussion following Jones 1966, p. 97.

49 Timmermann 2007.

50 Scadding 1967.

51 Treasure, personal communication.

52 Mountain 1986. On the history of cancer staging, see also Ménoret 2002.

53 Williams 1998.

54 Bailey 1984, p. 12.

55 Groot *et al.* 2007.

56 See also Link *et al.* 2005.

57 Craig, *The Last Freedom*, p. 16.

58 Craig, *The Last Freedom*, pp. 19–20.

59 Craig, *The Last Freedom*, p. 21. This attitude was representative of the dominant view in the profession: see Thatcher *et al.* 1995.

60  This is in line with the findings of The 2002.
61  Craig 1997.
62  Craig 1997, p. 33.
63  Craig 1997, p. 36. It is a sad fact, it seems, that cancer (and not exclusively cancer) patients often receive more treatment if they have personal connections or private insurance, or both, and not only in Britain. See also Diamond 1998.
64  Craig 1997, p. 37.
65  Craig 1997, p. 39.
66  *Ibid.*
67  *Ibid.*
68  Craig 1997, p. 40.
69  Craig 1997, p. 45.
70  Craig 1997, p. 46.
71  Craig 1997, p. 49.
72  *Ibid.*
73  Craig 1997, pp. 49–50.
74  Craig 1997, p. 50.
75  Craig 1997, p. 60.
76  *Ibid.*
77  Craig 1997, p. 61.
78  *Ibid.*
79  Craig 1997, p. 64.
80  Craig 1997, p. 74.
81  Craig 1997, p. 75.
82  Craig 1997, p. 77.
83  Craig 1997, p. 97.
84  Craig 1997, p. 108.
85  Wolpin *et al.* 2005.
86  Donnelly 2006.
87  This is illustrated by practitioner memoirs, for example Rosenberg and Barry 1992.
88  Maliski *et al.* 2003; Gauthier and Swigart 2003.
89  See the chapter by Keating and Cambrosio in this volume.
90  See also the chapter by Kutcher in this volume.

## Works Cited

Allison P. R. and Temple L. J. 1966, 'The Future of Thoracic Surgery', *Thorax*, 21, 99–103.

Abbey Smith R. 1957, 'The Results of Raising the Resectability Rate in Operations for Lung Carcinoma', *Thorax*, 12, 79–86.

Abbey Smith R. 1982, 'Development of Lung Surgery in the United Kingdom', *Thorax*, 37, 161–8.

Bailey A. 1984, 'The Epidemiology of Bronchial Carcinoma', in M. Bates (ed.) *Bronchial Carcinoma: An Integrated Approach to Diagnosis and Management*, Berlin: Springer, 11–22.

Barnes E. 2005, 'Caring and Curing: Paediatric Cancer Services since 1960', *European Journal of Cancer Care*, 14, 373–80.

Barnes E. 2007, 'Between Remission and Cure: Patients, Practitioners and the Transformation of Leukaemia in the Late Twentieth Century', *Chronic Illness*, 3, 253–64.

Barrett N. R. 1966, 'Introduction', in D. Teare and J. Fenning (eds) *Some Aspects of Carcinoma of the Bronchus and Other Malignant Diseases of the Lung. A Symposium Held at King Edward VII Hospital, Midhurst, July 4th and 5th, 1966*, Midhurst: King Edward VII Hospital, 1–4.

Bates M. 1984, 'Surgical Treatment', in M. Bates (ed.) *Bronchial Carcinoma: An Integrated Approach to Diagnosis and Management*, Berlin: Springer, 161–76.

Belcher J. R. 1956, 'Lobectomy for Bronchial Carcinoma', *Lancet*, i, 349–53.

Belcher J. R. 1966, 'Indications for Surgery and Choice of Operation', in D. Teare and J. Fenning (eds) *Some Aspects of Carcinoma of the Bronchus and Other Malignant Diseases of the Lung. A Symposium Held at King Edward VII Hospital, Midhurst, July 4th and 5th, 1966*, Midhurst: King Edward VII Hospital, 100–4.

Belcher J. R. 1983, 'Thirty Years of Surgery for Carcinoma of the Bronchus', *Thorax*, 38, 428–32.

Craig M. 1997, *The Last Freedom*, London: Hodder & Stoughton.

Diamond J. 1998, *C: Because Cowards Get Cancer Too ...*, London: Vermillion.

Donnelly R. 2006, *Cinderella Cancer: A Personal History of the Roy Castle Lung Cancer Foundation*, Liverpool: Bluecoat Press.

Doyal L., Green K., Irwin A., Russell D., Steward F., Williams R., Gee D. and Epstein S. S. 1983, *Cancer in Britain: The Politics of Prevention*, London & Sydney: Pluto Press.

Gauthier D. M. and Swigart V. A. 2003, 'The Contextual Nature of Decision Making Near the End of Life: Hospice Patients' Perspectives', *American Journal of Hospice and Palliative Care*, 20, 121–8.

Groot M., Derksen E. W. C., Crul B. J. P., Grol R. P. T. M. and Vernooij-Dassen M. J. F. J. 2007, 'Living on Borrowed Time: Experiences in Palliative Care', *Patient Education and Counselling*, 65, 381–6.

Harnett W. L. 1952, *A Survey of Cancer in London: Report of the Clinical Cancer Research Committee*, London: British Empire Cancer Campaign.

Hurt R. 1996, *The History of Cardiothoracic Surgery: From Early Times*, New York & London: Parthenon.

Ikeda S. 1970, 'Flexible Bronchofibrescope', *Annals of Otology, Rhinology and Laryngology*, 79, 916–19.

Jones A. 1966, 'Megavoltage Radiotherapy of Carcinoma of the Bronchus', in D. Teare and J. Fenning (eds) *Some Aspects of Carcinoma of the Bronchus and Other Malignant Diseases of the Lung. A Symposium Held at King Edward VII Hospital, Midhurst, July 4th and 5th, 1966*, Midhurst: King Edward VII Hospital, 95–9.

Krueger G. 2008, *Hope and Suffering: Children, Cancer, and the Paradox of Experimental Medicine*, Baltimore: Johns Hopkins University Press.

Lock S. A., Reynolds L. A. and Tansey E. M. (eds) 1998, *Ashes to Ashes: The History of Smoking and Health*, Amsterdam & Atlanta: Rodopi.

Link L. B., Robbins L., Mancuso C. A. and Charlson M. E. 2005, 'How Do Cancer Patients Choose Their Coping Strategies? A Qualitative Study', *Patient Education and Counselling*, 58, 96–103.

Maliski S. L., Sarna L., Evangelista L. and Padilla G. 2003, 'The Aftermath of Lung Cancer: Balancing the Good and the Bad', *Cancer Nursing*, 26, 237–44.

Martini N. 1993, 'Operable Lung Cancer', *CA: A Cancer Journal for Clinicians*, 43, 201–14.

Meade R. H. 1961, *A History of Thoracic Surgery*, Springfield, Ill: Charles C. Thomas.

Ménoret M. 2002, 'The Genesis of the Notion of Stages in Oncology: The French Permanent Cancer Survey (1943–1952)', *Social History of Medicine*, 15, 291–302.

Milstein B. B. 1991, 'The Cardiothoracic Society (Pete's Club) 1959 to 1989', *European Journal of Cardio-Thoracic Surgery*, 5, 339–45.

Mitchell D. M., Emerson C. J., Collyer J. and Collins J. V. 1980, 'Fibreoptic Broncho-scopy: Ten Years on (Occasional Review)', *British Medical Journal*, ii, 360–3.

Mountain C. F. 1986, 'Selecting Patients for Surgical Treatment of Lung Cancer', in C. F. Mountain and D. T. Carr (eds) *Lung Cancer: Current Status and Prospects for the Future*, Austin: University of Texas Press, 161–70.

Mountain C. F. 2000, 'The Evolution of the Surgical Treatment of Lung Cancer', *Chest Surgery Clinics of North America*, 10, 83–104.

Mueller C. B. 2002, *Evarts A. Graham: The Life, Lives, and Times of the Surgical Spirit of St. Louis*, Hamilton, Ontario: BC Decker.

Pack G. T. and Ariel I. M. 1955, 'A Half Century of Effort to Control Cancer; an Appraisal of the Problem and an Estimation of Accomplishments', in L. Davis (ed.) *Fifty Years of Surgical Progress 1905–1955*, Chicago: Franklin H. Martin Memorial Foundation, 59–161.

Pickstone J. V. 2007, 'Contested Cumulations: Configurations of Cancer Treatments through the Twentieth Century', *Bulletin of the History of Medicine*, 81, 164–96.

Rosenberg S. A. and Barry J. M. 1992, *The Transformed Cell: Unlocking the Mysteries of Cancer*, London: Chapmans.

Scadding J. G. 1967, 'Treatment of Bronchial Carcinoma (letter)', *Lancet*, i, 157.

Smart J. and Hilton G. 1956, 'Radiotherapy of the Lung: Results in a Selected Group of Cases', *Lancet*, i, 880–1.

Thatcher N., Ranson M., Lee S. M., Niven R. and Anderson H. 1995, 'Chemo-therapy in Non-Small Cell Lung Cancer', *Annals of Oncology*, 6, S83–S95.

Thatcher N. and Spiro S. (eds) 1994, *New Perspectives in Lung Cancer*, London: BMJ Publishing Group.

Timmermann C. 2007, 'As Depressing as it was Predictable? Lung Cancer, Clinical Trials, and the Medical Research Council in Postwar Britain', *Bulletin of the History of Medicine*, 81, 312–34.

The A.-M. 2002, *Palliative Care and Communication: Experiences in the Clinic*, Buckingham & Philadelphia: Open University Press.

Valier H. and Timmermann C. 2008, 'Clinical Trials and the Reorganization of Medical Research in post-Second World War Britain', *Medical History*, 52, 493–510.

Williams C. 1992, *Lung Cancer: The Facts*, 2nd ed., Oxford: Oxford University Press.

Williams C. 1998, 'Lung Cancer', in D. L. Morris, J. H. Kearsley, J. Kearsley and C. Williams (eds) *Cancer: A Comprehensive Clinical Guide*, London: Informa Health Care, 141–54.

Wolpin B. M., Chabner B. A., Lynch T. J. Jr. and Penson R. T. 2005, 'Learning to Cope: How Far Is Too Close?', *The Oncologist*, 10, 449–56.

# 4
# A Case Study in Human Experimentation: The Patient as Subject, Object and Victim

*Gerald Kutcher*

The case study has, in many respects, been replaced in post-World War II clinical trials by the investigation of cohorts. In medical reports and publications, the results of such studies are expressed through various statistical measures of survival, while stories of the individual patients almost never appear. While statistical measures of survival, which have been developed to a fine art, take centre stage in reporting the results of cancer clinical trials, complications are usually given towards the end of the paper, almost as an afterthought. And even here, the rate of complications is systematically underestimated favouring the physician's perspective over that of the patient.[1] These clinical trial representations are a poor proxy for the suffering of the patients.[2] Even physicians participating in clinical trials forfeit a good part of the individual character of their work, since in most randomised trials, they are denied access to accounts of the accumulating results of the trial before it has ended. In addition, the ethical ground for enrolling a patient in a clinical trial (the so-called principle of equipoise) is that the physician finds no reason to favour one arm of the trial over another based on a cohort of patients rather than the individual patient under consideration.[3] For problematic or notorious clinical trials (like the one we will recount below) the patient's story must not only find its way out from under the statistics, graphs and tables of the researchers, but it must also compete with the agendas of an increasing number of other actors (family members, ethical critics) who enter the story.[4]

Yet, the case study can provide a counter story to the grand narratives of medical researchers and has much to tell us about the anecdotal character of clinical trials, the suffering and courage of patients as well as ethical behaviour. Although reliance on cases goes back to Hippocrates, the case method began to decline in the nineteenth century with the development of scientific medicine that included pathological anatomy, histopathology and bacteriology. By the early twentieth century, however, physicians began introducing clinical cases, but primarily as a pedagogical tool.[5] With the introduction of the randomised clinical trial in the mid-twentieth century, the case method was fully displaced in medical research.

The method of cases has a much wider purview than medical practice. In Anglo-American law, cases are central to the use of precedent, unlike the continental method, which is based on fixed and stable rules.[6] And in the history of science, the study of cases has been notable, especially in the work of Thomas Kuhn who argued that normal science operates through paradigms. Investigators tackle new problems primarily through analogy with standard cases.[7] The case method (casuistry) has also been used for moral reasoning, beginning with Aristotle who understood that ethics could not be systematised as a science (*episteme*), but depended upon significant particulars, what he called *phronesis* or practical wisdom. But, in the mid-seventeenth century, Blaise Pascal put casuistry into disrepute, and with it accounts of morality based on the analysis of cases went into sharp decline. Recently, casuistry has made something of a return especially in the work of Albert Jonsen and Stephen Toulmin, who in the *Abuse of Casuistry* argue that the method of cases should be applied to problems in biomedical ethics.[8]

Finally, the case method has much to add to historical studies, especially as articulated by Carlo Ginzburg who suggests that the method is about the concrete and the individual and that they can produce knowledge that is 'indirect, presumptive, conjectural'.[9] Ginzburg assigns the source of the modern method to the Italian art critic Giovanni Morelli who was able to assess the attribution of a painting by focusing on minor details. He insisted that 'we should examine ... the most trivial details that would have been influenced least by the mannerisms of the artist's school: earlobes, fingernails, shapes of fingers and toes'.[10] Ginzburg points out the analogy with Morelli's contemporary Arthur Conan Doyle, whose detective Sherlock Holmes relies on minute evidence that is imperceptible to most people, but which provides clues that reveal the perpetrator of the crime. Morelli's belief that inadvertent little gestures reveal our character far more than formal gestures was later appropriated by Freud.[11] For Morelli, Holmes, Freud, and we might say, for historians as well, marginal data and discarded information are potentially significant for penetrating a case. It is also worth noting the importance of the analogy between clinical medicine and other case studies. Morelli, Doyle and Freud were all physicians and each recognised that the local, temporal and concrete were critical to unravelling their particular cases.

The subject of our case study, Maude Jacobs, was treated in the mid-1960s with total body radiation (TBI) for advanced cancer on a clinical trial of Dr. Eugene Saenger, a radiologist at the University of Cincinnati.[12] Saenger, who was under contract with the Department of Defence, used the patients as proxies for soldiers in order to study radiation effects from nuclear attack. For example, he used chemical measurements of urine samples from irradiated patients to try to develop a way of indicating the amount of radiation a soldier had received. Saenger claimed that TBI was intended to treat advanced cancer, while the military component of the study was only a secondary goal, a data gathering operation that had no effect on the clinical decisions. In spite of the concerns of some of his peers, the programme continued until public disclosures brought the studies to a close in early 1972. In the mid-1990s, Saenger's programme became public once again following a rash of press reports about unethical Cold War experimentation. Saenger's programme was investigated by a number of governmental committees including President Clinton's Advisory Committee on Human Radiation Experiments (ACHRE).[13] Although the committee reported in some detail on Saenger's experiments and tried to assess his ethical behaviour, it was silent on the individual stories of suffering of his patients.

In part, the absence of stories of individual patients in ACHRE's report was a consequence of the effort such a reconstruction would have entailed and the difficulty of doing so within the short time limit the committee was given. Indeed, the recovery and presentation of a case like Maude Jacobs, particularly a hospital case, presents numerous challenges. To begin with, the materials needed to reconstruct the case are very limited. There are of course surviving hospital records, but they have huge gaps in them, sometimes at the most critical junctures. For instance, there is nothing available for Maude's crucial hospitalisation following TBI. Also, the records are in terrible condition, since they were stored on microfiche and later photocopied in the 1990s. Even when they have been reasonably well preserved, many of the records are written in longhand. The handwriting is usually so poor that it is often difficult to reconstruct what an administrator, physician or other hospital worker meant to convey. Dates on notes are rare and initials (there are almost never signatures), if they appear at all, are almost always indecipherable. During later government hearings in 1994, the families of the patients complained bitterly about the state of the hospital records (see below). Since the hospital records were written, not by the patients but by hospital workers, the patients are presented in highly technical language; they appear more as medical objects than as human subjects. Of course, there is nothing unusual about poor hospital records; they are notoriously difficult to recover after a period of time and what is excavated is often incomplete and difficult to decipher since they are produced for use over a limited time period and are meant to be read by other professionals, certainly not the lay public, much less historians. Nevertheless, in spite of the difficulties, we can extract enough from the charts, Saenger's reports and publications, contemporary letters from Maude Jacobs to her aunt, later testimony of her children, and what I know about radiotherapy, as a long-term participant in radiation medicine, to imaginatively reconstruct some of Maude Jacobs' encounter with Saenger's clinical trial.[14]

The aim of this paper is to emphasise that patients play various roles in clinical trials and that such trials have an anecdotal character that is belied in traditional medical accounts. In the case of Maude Jacobs we consider her as subject, object and victim. In a full account, which we do not have the space to develop in this paper, each of these (and other) roles could be fully developed and would reveal, on the one hand, how the various roles are in tension and, on the other hand, how they also reinforce one another. In this paper, the primary emphasis is on Maude Jacobs' role as subject, while we only sketch her role as object and victim. In part, the decision to focus on her as subject, as I mentioned before, is to counter the standard stories of clinical trialists like Eugene Saenger. In the first two sections, we follow Maude Jacobs as subject and bring out her confrontation with a complex medical world as she tries to come to terms with her own illness. In the following shorter sections, we find her in very different roles; first as an object in the medical reports of Saenger and then as a victim in the narratives of her children over two decades later. Through the case method we can not only begin to appreciate the suffering of patients that is so poorly and inaccurately represented in medical reports, but also appreciate the complex and shifting relationships between patients and a multitude of hospital staff from medical researchers, to diagnosticians, clinicians, nurses, administrators and so on. In addition, through a detailed and layered recounting of the patient's story, the ethical nature of these various encounters is revealed in a way that escapes traditional narratives and the prescriptive ethics of modern medicine (e.g. whether a consent form has been signed).

## Maude Jacobs as subject: First encounter with cancer

Maude Jacobs noted a curious fact for her interviewer at the Cincinnati General Hospital (CGH) in July of 1964. She had fallen some 18 months previously and injured her right breast, arm and back. She felt a swollen area on the upper portion of her breast, which did not go away, but if anything, began to grow. Maude also recalled another defining moment, about one year after her fall, when her lower back and her right leg began to bother her. These two events were recorded on her admission note for CGH; the first denoted the onset of a problem, a moment that Maude identifies as the cause of her difficulties. The second is more crucial since that is what had brought Maude to CGH. It probably marked the moment when after a year pain was no longer a stranger, coming to visit every so often and then disappearing from consciousness, but a constant inhabitant of her body. Maude complained that the pain in her back had become more aggravated when she moved around or did any work at home. Perhaps it had already changed Maude's daily agenda, causing her to rest now and then, or to not take on certain tasks that she unconsciously realised might cause further discomfort. It was not just her back that bothered her; she told the interviewer that the pain extended along her right side from her thigh to her knee and all the way down to her ankle.[15]

On the following day, Maude went to the X-ray department to have 'pictures' taken. The X-ray report cautiously began by suggesting that she did not have metastatic disease in her lungs since 'the lung fields appear to be free of infiltrate'. The rest of the report was anything but positive and the bone survey was clearly foreboding: 'There is increased lucency of the spinous process of D-12, in addition there is thought to be an area of destruction involving the pedicle of D-12'.[16] It was not only her lower back that appeared to have metastatic lesions. On the lateral skull survey 'large destructive lesions' were noted in the frontal and parietal portion of her brain. Her right femur about half way down her leg had a 'large lytic metastatic lesion' and her pelvis was invaded with metastatic lesions on her right side, where her femur was attached to her hip, and also at the tip of her ileum. The report also noted an 'area of destruction involving the 6th left rib'.[17]

When Maude went for a surgical consultation, she probably told the surgeon a similar story about the onset and progression of her current difficulties, although we have no record of this conversation. The surgeon noted wide spread metastases on her X-ray films, but he was not sure whether the lesions in her pelvis and spine were cancerous or not. We know from the record that he palpated her right breast and was able to move the large lump rather freely since he stated that it appeared not to be attached to other structures. The surgeon also reported that he felt matted lymph nodes in her armpit, and these would likely represent the spread of her cancer and an ominous prognosis. When he pushed on her lower back Maude again felt sharp pain. And she must have been relieved when he began to test the reflexes in her extremities since he did not have to work very hard to elicit the responses he was seeking.

The surgeon's report evaluated Maude thus:

> Hard, irregular approximately 8–10 cm mass upper portion of right breast – non-tender and mobile, apparently not attached to the chest wall but nipple is inverted. Tenderness in R axilla c ... matted nodes anteriorly.[18]

In spite of the reported mobility of the lesion, which might suggest surgery, she was put on a course of chemotherapy. The surgeons had recommended chemotherapy since her tumour was too large and the cancer was too far advanced for surgery (combined with radiotherapy), although Saenger would later substitute total body radiation for chemotherapy.

We can only speculate about the role Maude played in the decisions regarding her treatment and what she understood and took in during those crucial days in July of 1964. It is possible that cancer was never openly discussed with Maude during her admission and chemotherapy in July, or later with Saenger's team. During the 1960s, it was still not unusual for physicians to avoid mentioning the word cancer to their patients. Cancer was then, as it is today, a fearful term and many people could not and did not confront it.[19] Since physicians were still viewed in high esteem (although that was beginning to change rather rapidly), they could tell patients what to do, what type of therapy they would undergo, even something as aggressive as chemotherapy, without raising the spectre of cancer.[20] Her physicians may have assumed that she did not want to know that she had cancer, and presented her treatments as a therapy or medication that would relieve her pain.

This is not an unreasonable assumption since Maude had ignored a serious level of discomfort for a year and a half when she came to the hospital. By then, the mass in her breast had literally grown to the size of a grapefruit and it had badly distorted her body. There are also other indications that suggest that Maude may have been in a state of denial. We have three letters she wrote to her aunt when she had returned home after her chemotherapy was completed. Maude never mentioned cancer, although all the letters are filled with stories about her debilitating pains and the difficulties that she was having in coping with her condition. In the last letter, Maude's complaint that 'I Don't think the Doctors noes what is wrong with me they didn't help me',[21] suggests that she was not willing to admit that she had cancer, and that it had spread throughout her body, although she was painfully aware of her debility.[22] One of her daughters, Sherry Brabant, in discussing Maude's letter at the ACHRE hearings 30 years later reached a similar conclusion that Maude 'was unsure of how sick she was'.[23]

Maude returned home following chemotherapy in the middle of August, happy to be with her children (there were three young ones) and with her God: 'I am so glad to be home with my kids ... I pray everything will fall in place'.[24] And later: 'Prayer help me God Brought me Through'.[25] But the pain continued and seemed if anything to be getting worse. In the first letter after her return, Maude complained: 'My arm is So Sore I Just Cant Bend My Elbow Sometimes'. By the second letter she despaired that 'I dont think I Ever Will walk good again'. Maude may not have wanted to know that she had cancer, but she appeared to be coming to terms with her own personal dilemma and the likelihood that she would never be the same. Her letters reveal how she could barely 'creep' around the house, how she held on to the furniture for support as she bravely attempted to do a little work – as she put it – and to take care of her children. In the third letter, there is a brief glimmer of hope as Maude proudly announced that she has been able to put on dress shoes recently. But reality immediately breaks in as Maude confesses that 'it take me few minutes to get on my feet'. Her letters probably remained one of her few outlets. For Maude, her life was her children, and she had to hide her distress from them and not admit to herself that they would be left alone sometime soon:

> When night Comes I So tired and Sore I have to go to Bed early. The Children wants me to Sit up watch TV with them ... I feel So Sorry for them Some times. But By the time the Day done I am Done for my Back hurts me So Bad it Just give away.[26]

## Maude Jacobs' encounter with TBI

In Maude's case report, Saenger stated that the chemotherapy treatment had markedly reduced the tumour in her breast, but that the metastatic lesions continued to grow.[27] The treatment was not bringing Maude any relief from her daily anguish. In spite of her pain and exhaustion, she continued to keep house and feed and take care of her children. By 2 November, three months after she completed chemotherapy, the pain was too much and Maude returned to CGH.

We have virtually no hospital records of the events that were to follow. We do not even know whom Maude talked with, and what options, if any, were discussed with her before she was given TBI. Yet a few of the records for the TBI study do remain and we also have indirect evidence that throws some light on what happened over the next month. We can surmise that Maude spoke with two of Saenger's collaborators, Dr Ben Friedman, a young internist, and Dr Harold Perry, a radiotherapist who had recently arrived from the Memorial Hospital in New York. Saenger was quite explicit in reports and publications that he did not prescribe therapy for any of the patients. He claimed that the decision to use TBI and the amount of dose was the joint decision of the internist and radiotherapist.[28] Friedman, who was interested in TBI as part of a bone marrow transplant programme, ran the experimental part of the study; it would have mattered very much to him which patients were chosen and what doses they would receive. Perry on the other hand was in charge of radiation treatments and he would have to deal with the consequences of TBI, especially if there were severe complications.

The family and other critics have questioned whether Maude gave written consent and how much she was informed about the nature of the TBI study and the possible outcomes. During the mid-1960s in the US, informed consent as a required procedure was still in its infancy and some institutions tried rather fitfully to implement some form of consent procedures, including the University of Cincinnati. However, the requirement of obtaining informed consent for clinical studies did not formally begin until 1966 (and then only for studies submitted to the National Institutes of Health for research support). Nevertheless, on the first point, whether there was written consent obtained for Maude Jacobs, we can be very confident: there was none. To begin with, there is no trace of it in the records that have been recovered. Although this alone does not prove that Maude did not sign a consent form, Saenger himself claimed that written consent began circa 1964–5 and certainly not before.[29] There are also blank consent forms in the archival record that suggest that written procedures began in May of 1965, actually somewhat before formal consent procedures were initiated at the University of Cincinnati.[30]

But even if Maude had signed a consent form, all we would know about her discussions with Friedman would be generalities. The form would have contained her signature and that of a witness to a statement that attested to a discussion of the 'value and purpose' of the therapy.[31] This consent process would not provide us with the crucial aspects of the encounter between Ben Friedman and Maude Jacobs, particularly how the issues were presented to Maude, the shades of meaning that were given to the words and the hope or despair that was created in the meeting. Indeed, during their meeting Maude was likely told little about the experimental nature of the treatment, the possible complications or even her own prognosis. In a letter to Saenger in 1971, Ben Friedman stated emphatically that after 1965 he went into great detail about TBI, that it was an experiment, and that he told the patients that they might not benefit from the treatment and so on.[32] Since Friedman would have had every reason to put the start of formal discussions with patients as early as possible, we can be fairly certain that nothing like that went on before 1965.

Friedman came into the consultation with Maude in November of 1964, during a period in which he was immersed in developing bone marrow transplants, and he would have been keen to continue to have patients enrolled in the TBI programme. In addition, he and Saenger were having difficulty finding suitable patients in which to escalate the dose of TBI, and he would have tried to enter Maude into the TBI study.[33] Maude, as we have mentioned, appeared to be in a state of denial and was clearly desperate to obtain some relief from her painful condition. Friedman would likely have quickly realised this, and he would have presented TBI as something that might help her (since TBI was meant to treat disseminated cancer). He would almost certainly not have discussed TBI as primarily an experiment, nor would he have emphasised that she might have depression of her blood cells due to damage to her bone marrow and how that could lead to bleeding, infections and even death. Maude, on the other hand, would likely have listened for anything that would convey some hope, and she probably would have tuned-out any references to a 'study' that Friedman may have briefly alluded to.

Whatever was said during Maude's encounter with Friedman (and later Perry), the evidence suggests that she emerged understanding little, if anything, more about her medical condition. She also clearly knew nothing of her role as a proxy for soldiers. During testimony before an ACHRE hearing in 1994, her daughter Lillian claimed that following TBI she never left Maude's side and that Maude never mentioned that she was part of a study. Lillian also stated that during public disclosures about Saenger's experiments in the early 1970s, she had argued with others that her mother 'had nothing to do with it', that is, with the TBI experiments.[34]

Four days before TBI was initiated, Maude had blood drawn for immunological studies to be carried out at Fort Knox and for haematological, chromosome, and other studies at CGH. These tests were repeated twice and a urine analysis was performed prior to her treatment.[35] With these assessments Maude had been calibrated as a human dosimeter for Saenger's study. According to the TBI protocol, the investigators were expected to proceed only: if Maude's haemogram was stable,[36] if she was in a generally good nutritional state, if she had normal kidney function, and if she had disseminated disease.[37] With these conditions presumably met, Maude went on to TBI on 7 November.[38]

Maude Jacobs, no doubt nervous and uncertain about her future, must have arisen early that day since her therapy was scheduled for the morning.[39] When she arrived in radiation therapy, she would have been told to change into a hospital gown and to wait to be taken to the therapy machine. She may have met Perry but more likely she was passed on to the radiation technologists who were in front of the room where she was to be treated. Maude would have been led into the room down a narrow corridor that opened out onto a brightly lit and rather forbidding space filled with objects whose purposes she could not possibly appreciate. There would have been thick, heavy (lead) blocks of different sizes and shapes scattered about, white masks and body moulds covered with strange markings, as well as assorted metallic and plastic objects. The room would have seemed foreboding since there were no windows or natural lights. There was only the oppressive and austere feeling of being in a cell or a dungeon.[40] We, of course, do not know whether Maude had these very responses, or whether she noticed any of the devices in the room, but she could not have missed the large machine dominating the room with its enormous bulbous head with 'El Dorado' written across its front.[41] She would have realised right away that it was the therapy machine, the Cobalt unit that would give her the treatment. She was taken to the far side of the room, to a platform with a chair on it near one of the walls. The technologists would likely have had to help Maude up onto the platform because of the pain in her back and her right leg. They would also probably have explained to her at this point that this was where she would get her treatment. The chair that she finally sat on was facing one of the walls and not the Cobalt machine, which was off to her side. If she had peered over her shoulder she might have seen it moving in a large arc as one of the technologists stood beside it watching it move. As it came to rest, what had been the lower portion of the bulbous head was now facing towards her with its front open like a large yawning mouth.

The technologists would have asked Maude to lean forward in her chair, pulling her legs up towards her like a foetus. They would have said that it was important to hold still during the treatment so that her whole body would remain within the field of radiation. During these moments, one of the technologists would have taped some small objects to her pelvis, her chest, and her head, first on one side and then on the other. If she asked what they were doing, she would have been told that the objects were little measuring devices that were used to record her treatment.

Suddenly, the room lights would have been extinguished except for a bright swathe of light that she would have felt shining on her. If she turned towards the wall she would have seen her body framed in a large square of light, barely enclosing her head and feet. But before Maude could have mused for very long, the room lights would have come back on, and the technologists would have started towards the entry way as they instructed her to hold her position until the treatment was over. That would take almost half an hour, and she would also have been informed that they would be watching her and that she could call out if she needed any assistance. After what must have seemed an eternity, the technologists would have come back into the room, and they would have rotated the platform and repeated the treatment. We know nothing of what Maude experienced during that long wait. According to the testimony of some patients who have undergone similar treatments, she would have had no particular reactions and she should have felt nothing at all, no pain, burning or other physical sensations.[42] Maude may have felt a bit woozy during the treatment, especially if she sat very still and did not exercise her muscles to help her blood circulate. She may have prayed or communed with God during that endless wait; certainly He had brought her through worse moments.[43]

At the end of the session the situation changed radically; as the technologists helped Maude down from the platform, she began to feel a bit woozy. Almost immediately, she was carried off by a severe bout of nausea and vomiting. The case history is quite explicit on this point: 'At the termination of treatment, the patient had severe vomiting which continued throughout the next 24 hours in spite of intramuscular compazine'.[44] Nothing could stop the vomiting nor alleviate the violent paroxysms that seized her over and over again. Even after the worst had passed, and the seizures had become attenuated and less frequent, she must have felt severely depleted and exhausted from her bouts of vomiting, and she must have been very frightened.

Maude could not get out from under her worsened state. The radiation had made her terribly ill, and it had not brought any relief from the constant pain. Her right leg was if anything worse than before. Maude fell into a state of lethargy and depression. She had no appetite. If she forced down some food there was every possibility that she would become nauseous again and begin another round of vomiting. Her caloric intake had to be near minimal, and she was probably often dehydrated, with headaches and fever. Maude lasted four days after treatment with TBI before she was readmitted to CGH.[45]

Maude's final days were terrible. Within a week of the TBI, her lungs became filled with infiltrates. One lobe had collapsed altogether and she was near to choking to death. Her heart was galloping out of control trying to pump more and more oxygen into her system. Her immune system was on the verge of collapse. Her white cell count was falling rapidly, as were her platelets. And she continued to vomit. Maude's symptoms were those of acute radiation sickness, something a soldier in nuclear war might experience. Antibiotics were started to help her immune system fight off opportunistic infections. Heparin was administered for her heart condition, even though her platelet counts were falling and she might begin to bleed internally.[46] Sherry, one of her young children at the time, remembered that Maude 'was very depressed' and 'out of her head quite a bit of the 25 days she survived'.[47] By the third week following TBI, Maude was still vomiting, but now it was filled with dark pellets, an ugly and terrifying omen.[48] Her suffering finally ended. The death certificate reads that Maude Jacobs, a white female housewife, died on December 2 1964: Cause of death – 'Carcinoma of the Breast'.[49]

## Maude Jacobs objectified: From cancer patient to cohort member

The statement that Maude's death was due to breast cancer implies that something approaching the natural history of the disease had led to her demise. The effects of radiation on her system and its role in her death were not mentioned at all. Although in Saenger's reports to the Department of Defence, Maude had been constructed as a surrogate soldier, at CGH, she was simply a cancer patient. As Maude passed from Friedman to Perry to the technologists at the Cobalt machine, to her home, and then back to the hospital, she gradually and successively passed from the role as surrogate soldier and human dosimeter, to another as cancer sufferer and patient. To Friedman, she was someone he wanted to enrol in his study. To Perry she was predominately a cancer patient who needed treatment for her painful metastases. No doubt she was also a potential study patient who he had discussed with Friedman. He also had to make sure the treatment followed the protocol and he had to follow her responses to TBI. But for Perry she had most likely more than partially shed her experimental role. For the technologists at the Cobalt-60 machine, Maude was a sick patient who would receive TBI. They understood she was part of a study, but they would have known little or nothing about the details beyond how they were expected to carry out the treatment. It would have been quite unusual for the technologists to play any further role in the TBI project; they would neither have read nor have contributed to the various meetings, reports, and publications. By the time Maude returned to the hospital all visible traces of her role as a patient on an experiment were virtually obliterated for those who came in contact with her, except perhaps for the detailed blood studies that continued every few days. When she died she was a person suffering from cancer whose medical problems due to cancer were so enmeshed in those that were due to therapy that no one could or would try to pull them apart. Nor was her iatrogenic disease so very different from that of a breast cancer patient who had undergone more routine chemotherapy instead of TBI. They also could have had bouts of lethargy, nausea, vomiting, and opportunistic infections similar to Maude. With either therapy, the patients could have suffered substantially from a combination of their disease and the treatment.

If Maude Jacobs was construed for the Defence Department as a surrogate soldier, and for CGH as cancer sufferer and patient, she played at least two more roles after her death, since she was considered a patient on a cancer study according to Saenger in the early 1970s, and as a victim of the Cold War according to her children in the 1990s. In a 1973 paper entitled 'Whole Body and Partial Body Radiotherapy of Advanced Cancer', Saenger subsumed Maude into various cohorts of patients that he claimed had received TBI as part of a study, not for military purposes, but for treating advanced cancers. Maude was eligible for the study since, according to Saenger, she had 'advanced cancer for which cure could not be anticipated'.[50] She was one of 15 breast cancer patients who had undergone TBI and whose survival Saenger reported 'appears somewhat better than that of the patients treated solely by estrogens and androgens but not quite as good as the group treated with 5-flourouracil'.[51] Indeed, Saenger claimed that the median survival following TBI in this group was 445 days, a result that Saenger felt important enough to report as a contribution to the state of knowledge in cancer care. In addition, Saenger would claim limited suffering of the patients not different from chemotherapy:

> The analysis of our 88 treated patients show that ... in only 4 patients (4 per cent) were the nausea and vomiting of a severe nature. These symptoms are no greater than found after surgery or after treatment with cancer chemotherapy drugs.[52]

Although Maude may seem to be lost in the tables and graphs that Saenger presented, we are in a position to find her in the mass of figures. Saenger wrote in the paper 'one can identify 8 cases in which there is a possibility of the therapy contributing to the mortality'. Maude was part of that group as one of the two who had 'extensive previous chemotherapy'. She can be found in table IV where patient 45 is paired with a survival of 138 days following diagnosis and 25 days following 150 rads of TBI.[53] We can also locate her in one of the survival curves, where she is the fourth circle from the origin, one of the poorest survivors on the graph.[54] Maude revealed yet another trace, as one of the 19 patients who died within 20–60 days.[55] Saenger compared those 19 patients with another group of cancer patients who he did not treat with TBI and who also survived between 20 and 60 days. Using a statistical measure to compare the two groups, Saenger argued that there is no survival difference between them. He concluded that 'in other patients described [that is, Maude and her 18 companions], the effect of whole and partial body radiation therapy was less important in contributing to death than was the extent of disease in these patients'.[56] According to Saenger, Maude had died from her cancer and not from bone marrow syndrome; her death certificate got it right.

### Maude Jacobs as victim

While Saenger saw Maude as one of a cohort of patients who had died predominantly from cancer, her children came to consider her not as a cancer sufferer but a relatively healthy individual who was a victim of Cold War experiments. Although her daughter Sherry acknowledged in a letter to ACHRE that Maude was in pain, her difficulties were presented in the context of someone who 'cleaned and cooked and cared for me and my two sisters'.[57] Maude 'could possibly have gone into remission' and may have been saved since 'a cure could have been found'. Maude had not died of cancer, but was 'murdered' by Saenger.[58]

The children also spoke of the devastation that the deaths of their parents brought to them and the rest of the family. 'These "Doctors" left my Mother with no job skills, to raise a grandchild' and she died a 'broken women after my Father's premature death'.[59] For Sherry Brabant, Saenger had condemned her 'to hell' since she and her three sisters were forced to go to an orphanage and they were separated after Maude's death.[60] The possibility of coming to terms with death, and the potential for healing over time was obliterated for her family. Only pain, anger and frustration remained. During and after the ACHRE hearings old wounds had been opened up and discussed in the new language of informed consent and human rights. The children could only wonder in exasperation and in cold fury how such experiments on patients were possible. Many of them could only see conspiracy and victimisation.

The children were also disturbed by the difficulty of obtaining Maude's hospital records. Even when Sherry Brabant finally received them, she was angry and frustrated when she saw how little information they contained and how hard the records were to decipher. During her appearance at public hearings held by ACHRE in Cincinnati in 1994, Sherry and other family members expressed their rage at the condition and paucity of the records they received. Sherry asked the ACHRE staff, how was it possible that so much was missing from her mother's hospital chart? 'Things like doctors' notes and nurses' notes, the kind that you would have on the clipboard at the bottom of the bed' were all missing.[61] The photocopies she had were in very bad condition. There were many places where they could not be read because blank paper had been photocopied over the underlying writing. This may have occurred when the microfiche copies were made more than 25 years ago or during photocopying in 1994.

Yet the charts should not be judged entirely by our own special requirements. The lack of signatures and dates, the hand-written notes that are almost impossible for us to understand, work within the hospital environment. Medical workers are quite adept at searching through hospital records and getting to the information they need. They can rapidly decipher the scribbles and scrawls of their colleagues and extract the one or two nuggets that they are searching for. It could not be otherwise, since only a fraction of the knowledge on a patient is recorded explicitly in the charts. Behind the cryptic notes, there is a world of tacit medical knowledge that is shared among medical workers and provides them with a framework in which they can extract the medical essence from those cryptic notes.[62]

The patients and their children were strangers to that framework and all that goes on in a hospital. Maude, we remember, experienced some of the strangeness of being a patient. She had to participate as an outsider in a foreign ritual using strange objects and choreographed according to rules that only the hospital workers knew. The entire process, except for the medical effects it was supposed to have, would have been opaque to Maude's gaze. The children also experienced some of that strangeness as they tried to reconstruct the history of their mother from the hospital records – and they have found little comfort. Victimisation was a powerful and double-edged metaphor; it gave meaning to her suffering, but it also may have distanced them by taking away some of her humanity. For example, Sherry's story of her mother's victimisation took on such proportions, that in a letter to ACHRE she wondered, among other things, whether Maude had been buried in Baltimore Pike Cemetery with her organs removed.[63]

## Conclusions

The case method, with its specificity and emphasis on minor details, reveals aspects of the trials and tribulations of the patients that are entirely absent from reports of the medical researchers. Saenger's publications are typical examples of such medical research reports with their emphasis on positive outcomes, that is, on the survival value of the therapy, while the complications of the treatment are given only passing mention with little detail, all of it conveyed in dispassionate terms.[64] In contrast, through the case study approach we can begin to appreciate the suffering of the patients and give some substance to the so-called 'side effects' (or we might say in a different context, the 'collateral damage') of the treatment. It is not only the nature of the response to therapy that emerges from the record, but also the bewildering and frightening nature of the whole therapeutic process. In many ways Maude was a typical participant in that she saw her treatment as a ritual whose language, movements and objects she could barely understand. Although today patients (at least middle-class, educated patients), as consumers of medicine, are much better informed than they were in the 1960s, it is hard to imagine that they find themselves any less alienated when experiencing a radiation treatment.[65]

Through the case study, the layered and varied relationships between patients and medical staff also emerge as we come to appreciate Maude's constantly changing role from a participant on an experimental trial to a clinical patient. This more nuanced picture replaces the simple dichotomy of tension between the experimental subject and clinical patient so typical of writings on clinical trials. The case study also reveals much about the character of the participants that again is wholly absent from clinical trial reports. In Maude Jacobs' case, we can not only appreciate her difficulty in confronting a terrifying disease like cancer, but we also can admire her ability to carry on and care for her children in the face of her debilitating condition. And the case study provides us with insights into the relationship between patient and physician that cannot be extracted from medical publications and informed consent documents. Minor details – how the physician presents the study, how its benefits are pitched, how much detail about the complications are revealed, the type of questions the patient asks and how the physician responds – are crucial to the outcome of the encounter and reveal much about the ethical behaviour of the participants. In addition, the pressure physicians feel to enrol patients in their study and the patient's search for a better therapy come together in an encounter that signed informed consent statements tell us nothing about. Finally, a case study of notorious clinical trials like Saenger's makes it possible to more fully appreciate the controversies that surround such studies. In Maude's case, the family and critics constructed her as a victim ground up in the medical machine. Yet, the case study makes room for a richer interpretation that provides a narrative of Maude's strength of character.

## Acknowledgements

Parts of this article have been published by the author in *Contested Medicine: Cancer Research and the Military*, © 2009 by The University of Chicago. All rights reserved. I would like to thank Simon Schaffer for comments on an earlier version, Carsten Timmermann for his support and suggestions and Marilynn Desmond and Elizabeth Toon for their careful reading and detailed comments and suggestions.

## Notes

1    Complications are given as the ratio of the number of patients with a complication to those entering the trial rather than to those surviving.
2    For clinical trials see Hill 1951, Pocock 1983, Meinert 1986, Geehan and Lemak 1994 and Berg 1997. For a typical example of how clinical trial results are presented see the classic paper by Fisher *et al.* 1977.
3    See e.g. Fried 1974 and Hellman and Hellman 1991.
4    The stories of patients' battles with disease, particularly cancer, are ubiquitous. These narratives are usually told primarily, if not entirely from the patient's perspective. There are, however, histories in which the viewpoints of both patients and medical elites co-exist. For example, Wailoo 2001 tells us about the slow process by which the invisible suffering has been made visible while he does not let the search for a cure for sickle cell anaemia eclipse the plight of black Americans.
5    For a nuanced discussion of cases, see Forrester 1996.
6    *Ibid.*
7    Kuhn 1962.
8    Jonsen and Toulmin 1988 and Toulmin 1982.
9    Ginzburg 1989, p. 106.
10   Ginzburg 1989, p. 97.
11   Ginzburg 1989, pp. 97–9.

12 For a fuller discussion of the Saenger case and its value in understanding post-WWII research practices and clinical conduct see Kutcher 2009. For overtly ethical stances, see ACHRE 1996 and Stephens 2002, who also presents an alternate discussion of Maude Jacobs.

13 ACHRE 1996, pp. 239–48.

14 The author spent most of his career in Radiation Medicine as Chief of the Clinical Physics Service at Memorial Sloan-Kettering Cancer Centre in New York.

15 Admission note to CGH July 16, 1964, IND 102194-F.

16 X-ray Report July 17, 1964, IND 102194-F.

17 *Ibid.*

18 Female Surgery Consult ND, IND 102194-F.

19 Patterson 1987, pp. 167–70.

20 For medical paternalism see e.g. Cooter 2000, pp. 451–68, Faden and Beauchamp 1986, Rothman 1991 and Wolpe 1998, pp. 38–59. For attacks on medicine beginning in the 1960s see e.g. Illich 1976, Kennedy 1981, pp. 243–8, McKeown 1976.

21 I am using the spelling and grammar in Maude's letters. There would be no purpose in using *sic.*

22 Jacobs, third letter to Aunt, 'Just a few lines to answer your letter ...', ND, IND 102194-F.

23 Brabant, notes to ACHRE, 'Points to be made ...', ND, IND 102194-F.

24 Jacobs, first letter to Aunt, 'Will take time to ...', ND, IND 102194-F.

25 Jacobs, second letter to Aunt, 'Just a few lines to let you no ...', ND, IND 102194-F.

26 Jacobs, third letter to Aunt, 'Just a few lines to answer your letter ...', ND, IND 102194-F.

27 Saenger E., Feb. 1960–April 1966, 'Metabolic Changes in Humans Following Total Body Radiation', *Progress Report*, Case History for Patient 45. DOD 042994-A;7/16. From now on cited as Saenger, Progress Report.

28 Saenger *et al.* 1973, p. 672.

29 R. Neumann *et al.*, 15 Sept., 1994, Interview with Dr. Eugene Saenger, Cincinnati, ACHRE Interview Project, p. 62.

30 The institution began formal consent procedures following US Public Health Service requirements. Kutcher 2009.

31 Saenger, Progress Report, Fig. 1.

32 Friedman to Saenger, Dec. 15, 1971, 'In regard to your phone call of Dec. 13, 1971 ...'. DOD 042994-A;11/16.

33 Kutcher 2009.

34 Testimony of Lillian Pagano, 21 Oct. 1994, ACHRE Public Hearing, Cincinnati, p. 84, http://www.seas.gov.edu/nsarchive/radiation/dir/ mstreet/comeet/pm01/pl1tran.txt/ From now on cited as ACHRE Cincinnati Meeting.

35 Reporting form for TBI Study, 'Blood, Urine', ND, CORP 091394-B.

36 Actually her haematological profile was zero, the normal value for a healthy individual.

37 Saenger, Progress Report, p. 3.

38 Saenger Progress Report, Case History for Patient 45.

39 The description in this section is based on the therapy protocol in *ibid*, pp. 3–5 and pp. 66–70 and the author's experience in radiation medicine.

40 For a discussion of radiation therapy machines, accessories, and treatment rooms, see Johns and Cunningham 1983.
41 Kereiakes *et al.* 1972, p. 652.
42 Based on the author's observation of patients undergoing total body treatments.
43 The TBI protocol is described in Saenger, Progress Report, pp. 4–18.
44 Saenger, Progress Report, Case History for Patient 45.
45 *Ibid.*
46 *Ibid.*
47 Testimony of Sherry Brabant, ACHRE Cincinnati Meeting.
48 Saenger, Progress Report, Case History for Patient 45.
49 Death Certificate for Maude Jacobs, 2 Dec. 1964, IND 102194-F.
50 Saenger *et al.* 1973, p. 671.
51 Saenger *et al.* 1973, p. 675.
52 Saenger *et al.* 1973, p. 679.
53 *Ibid.*
54 Saenger *et al.* 1973, p. 678.
55 *Ibid.*
56 Saenger *et al.* 1973, p. 679.
57 Brabant Testimony, ACHRE Cincinnati Meeting.
58 Brabant notes to ACHRE, 'Points to be made ...', ND, IND 102194-F.
59 *Ibid.*
60 *Ibid.*
61 Brabant Testimony, ACHRE Cincinnati Meeting.
62 See Risse 1999, pp. 155–86 and Timmermans and Berg 1997.
63 Brabant Testimony, ACHRE Cincinnati Meeting.
64 Saenger's DOD reports on the contrary are remarkable for its detailed descriptions of the complications of each of the patients. However, these individual stories are absent from the published record. See Saenger 1967 and Saenger *et al.* 1973.
65 For a discussion of the use of clinical trials as a way of coping with cancer see, Löwy 1996, pp. 73–83.

## Works Cited

### Primary Sources

ACHRE Archives, Record Group 42, National Archives II, 8601 Adelphi Road, College Park, Maryland, 20740-6000, US. The sources contain material from the Department of Defence (DOD), the Department of Energy (DOE), corporate (CORP) and independent sources (IND). A DOD entry, for example, would appear as DOD 042994-A;7/16, which refers to source, date it was received, batch code, file number, and the total number of files in the batch.
Fisher B., Glass A., Redmond C., Fisher E. R., Barton B., Such E., Carbone P., Economou S., Foster R., Frelick R., Lerner H., Levitt M., Margolese R., MacFarlane J., Plotkin D., Shibata H. and Volk H. 1977, 'L-Phenylalanine Mustard (L-PAM) in the Management of Primary Breast Cancer: An Update of Earlier Findings and a Comparison with Those Utilising L-PAM Plus 5-Fluorouracil (5–FU)', *Cancer*, 39, 2883–903.

Kereiakes J. G., Van de Riet W., Born C., Ewing C., Silberstein E. and Saenger E. L. 1972, 'Active Bone-Marrow Dose Related Hematological Changes in Whole-Body and Partial-Body Co-60 Gamma Radiation Exposures', *Radiology*, 103, 651–6.

Saenger E. L. 1967, 'Effects of Total- and Partial-Body Therapeutic Irradiation in Man', *Proceedings of 1ˢᵗ International Symposium on Biological Interpretation of Dose from Accelerator-Produced Radiation,* University of California, Berkeley, March 13–16, 113–27.

Saenger E. L., Silberstein E. B., Aron B., Horwitz H., Kereiakes J. G., Bahr G. K., Perry H. and Friedman B. I. 1973, 'Whole Body and Partial Body Radiotherapy of Advanced Cancer', *Am. J. Roentgenol. & Rad. Therapy*, 117(3), 670–85.

## Secondary sources

ACHRE 1996, *Final Report of Advisory Committee on Human Radiation Experiments,* Oxford: Oxford University Press.

Berg M. 1997, *Rationalizing Medical Work: Decision-Support Techniques and Medical Practices,* Cambridge, MA: MIT Press.

Cooter R. 2000, 'The Ethical Body', in R. Cooter and J. Pickstone (eds) *Medicine in the Twentieth Century,* Australia: Harwood Academic Publishers, 451–68.

Faden R. and Beauchamp T. 1986, *A History and Theory of Informed Consent,* New York: Oxford University Press.

Forrester J. 1996, 'If p, then What? Thinking in Cases', *History of the Human Sciences*, 91, 1–25.

Fried C. 1974, *Medical Experimentation: Personal Integrity and Social Policy,* Amsterdam: North Holland Publishing.

Gehan E. and Lemak N. 1994, *Statistics in Medical Research: Developments in Clinical Trials,* New York: Plenum Medical Book Company.

Ginzburg C. 1989, 'Clues: Roots of an Evidential Paradigm', in C. Ginzburg, *Clues, Myths, and the Historical Method,* Baltimore: Johns Hopkins University Press, 96–125.

Hellman S. and Hellman D. 1991, 'Of Mice But Not Men: Problems of the Randomized Clinical Trial', *N. Engl. J. Med.*, 324, 1565–89.

Hill A. B. 1951, 'The Clinical Trial', *Brit. Med. Bull.*, 7, 278–82.

Illich I. 1976, *Medical Nemesis: The Expropriation of Health,* London: Calder and Boyars.

Johns H. and Cunningham J. 1983, *The Physics of Radiology,* 4ᵗʰ edition, Springfield: Charles Thomas.

Jonsen A. and Toulmin S. 1988, *The Abuse of Casuistry: A History of Moral Reasoning,* Berkeley, CA: University of California Press.

Kennedy I. 1981, *The Unmasking of Medicine,* London: George Allen and Unwin.

Kuhn T. 1962, *The Structure of Scientific Revolutions,* Chicago: University of Chicago Press.

Kutcher G. 2009, *Contested Medicine: Cancer Research and the Military,* Chicago: University of Chicago Press.

Löwy I. 1996, *Between Bench and Bedside: Science, Healing and Interleukin-2,* Cambridge, MA: Harvard University Press.

McKeown T. 1976, *The Modern Rise of Population,* London: Edward Arnold.

Meinert C. 1986, *Clinical Trials: Design, Conduct and Analysis*, Oxford: Oxford University Press.

Patterson J. 1987, *The Dread Disease: Cancer and Modern American Culture*, Cambridge, MA: Harvard University Press.

Pocock S. 1983, *Clinical Trials: A Practical Approach*, Chichester: John Wiley.

Rothman D. 1991, *Strangers at the Bedside: A History of How Law and Bioethics Transformed Medical Decision Making*, New York: Basic Books.

Risse G. 1999, *Mending Bodies, Saving Souls: A History of Hospitals*, New York: Oxford University Press.

Stephens M. 2002, *The Treatment: The Story of Those Who Died in the Cincinnati Radiation Experiments*, Durham, NC: Duke University Press.

Timmermans S. and Berg M. 1997, 'Standardization in Action: Achieving Local Universality through Medical Protocols', *Social Studies of Science*, 27, 273–305.

Toulmin S. 1982, 'How Medicine Saved the Life of Ethics', *Perspectives in Biology and Medicine*, 25, 736–50.

Wailoo K. 2001, *Dying in the City of the Blues: Sickle Cell Anemia and the Politics of Race and Health*, Chapel Hill: The University of North Carolina Press.

Wolpe P. 1998, 'The Triumph of Autonomy in American Bioethics: A Sociological View', in R. DeVries and J. Subedi (eds) *Bioethics and Society: Constructing the Ethical Enterprise*, Upper Saddle River, NJ: Prentice Hall.

# 5

# Captain Chemo and Mr Wiggly: Patient Information for Children with Cancer in the Late Twentieth Century*

*Emm Barnes Johnstone*

This is a story about stories, told to and by children with cancer, over the past 40 years in Britain. Asking why particular stories were narrated to this group of patients as a distinct population sparked a historical inquiry into why children came to be seen as having special needs when undergoing medical treatment.[1] This also prompted an exploration of how patients have been informed about their conditions and treatments.

Patient information literature has escaped historical attention. The genre sits awkwardly between illness narratives and health education material. Yet it provides a window on developments in patient experience and doctor–patient encounters and expectations. Child patients in particular have not received attention. Historians have certainly attended to the changing ways in which childhood has been seen by professionals in health, education, legal systems and by parents. But experiences of being viewed by professionals and parents have been less intensively studied. Children's attempts to resist and recast the meaning of illness are hard to locate and have not yet been fully integrated into the historical study of child health.

This article outlines what patient information is and describes its utility for the historian. It goes on to assess the emergence of concerns for the psychological welfare of sick children in the post-war years. The article then traces two distinct histories and how they relate to one another: the emergence of childhood cancer as a major problem and the use of cartoon booklets as a means of communication between patients and doctors. Evolving representations of chemotherapy serve as examples through which to understand wider changes in information provision and conceptualisations of the needs of sick children. The article concludes by demonstrating that, through the study of such sources, children's contributions to cultural dialogues about the meaning of illness and patienthood may be recovered.

## Patient information

We are concerned here with materials produced over the past 40 years for children with cancer.[2] The term 'children' is taken to denote people under the age of 18.[3] Patient information for children with other chronic health problems has become increasingly available over the past decade. Childhood cancer serves as a particularly good case study through which to explore the functions of, and value to the historian of, this kind of source material.[4] This is for a number of reasons.

First, more information has been produced for cancer than for any other childhood disease. This may be because there has been extra charitable funding for children's cancer per child affected than for any other disease. In addition, cancer has a distinctive narrative arc, in that it *can* be cured for some patients, but cannot always be cured and may result in death. Information is designed to help children and their carers get through a finite course of treatment with the expectation that normal life will subsequently resume. Second, information was first widely distributed to children with cancer in the 1970s and to parents even earlier. This makes it possible to examine changes over time in what has been expected from patient information. Attending to how materials have varied by age of anticipated reader suggests the extent to which people under the age of 18 have been assumed, by medical professionals and themselves, to constitute a homogenous group.

Third, while the cultural meaning of cancer has been the subject of many studies, the ways in which children understand it have largely been ignored outside the psychological and anthropological communities.[5] Young audiences have been offered, and have re-mediated, ways of understanding 'the dread disease' that are markedly different from those among adults. They have shared their visions of cancer widely and thus may have affected the broader cultural meaning of the disease.

How does patient information relate to other sources? First, it is ephemeral. Prior to moves within the National Health Service (NHS) in the 1990s to improve communication between service users and providers, materials for patients attending hospital for procedures or medication were locally produced, and publication details were not recorded. They varied widely between hospitals and even departments within one hospital. As for the literary form of these sources, materials produced for children have been markedly different from those generated for older patients; there are more similarities between those produced for children with different diseases than materials for children and adults with the same disease. Leaflets for adults typically resembled, and resemble, basic medical guidebooks or 'self-help' manuals designed to impart coping skills: readers do not, as it were, meet fellow-patients in these materials.[6] One might speculate that adults are familiar with DIY manuals, and that this is the borrowed form for guidance on health. It is also more likely that adults have already come into contact with other ill people and thus have a stock of stories through which to filter in their own experience.

The information given to ill children, however, adopts and retains a different form. Materials follow a patient's chronological path through medical services and psychological adjustment. Who this child is, and the nature of his or her relationship to the reader, has changed substantially over the 40-year history of the genre. But it is this storytelling form that makes patient information materials aimed at children revealing for the historian. They present professionals' expectations *and* children's experiences of being treated.

Patient information can be seen as closely related to four narrative traditions. First, there are case-studies as presented in medical literature to illustrate treatment innovations and outcomes. Second, we have information about treatment for new patients. Third, there are personal stories recounted in the media as a means of bringing medical conditions alive for the public. Finally, patient information for children also draws on traditions within children's literature. Thus medically produced information came to be 'storified' for children, though rarely so for adults. The first narrative tradition, that of the case-study or case story in medical teaching, will not be discussed here. While a doctor's narration of a case is designed to inform colleagues, a patient's story, when it takes the typical form of a memoir, denies that the ultimate significance or meaning of a symptom is to be decided by the doctor.[7] Illness narratives that have been explicitly constructed for patients are neither case stories nor patient memoirs. They read and serve as case-studies rather than memoirs; the tutees are patients not doctors. This is the most distinctive feature of the stories presented in patient information.

The second narrative tradition is that of using the experiences of former patients as prognostic indicators. The experiences of children diagnosed with cancer have directly shaped the information given to families. From the early 1960s, British doctors and parents recorded information exchanges from the time of diagnosis onwards.[8] Limits of the range of possible outcomes for a child were defined on grounds of past patients' experiences and, in the 1970s, this information was collated in select treatment centres. Local sources were then collected by panels appointed by professional bodies, and replaced by centrally produced materials.[9]

The third narrative tradition is that of the human interest story, as seen in memoirs and news. In recent years, the history of the human interest story has come under historical scrutiny.[10] The print media, radio and television traditionally break news about medical research by connecting it to personal stories of people affected with an associated condition. Patient information can be seen to have taken a very similar course over the past 40 years: individuals' stories illustrate the state of medical knowledge. Additionally, children have seen that the stories of others can attract attention for specific reasons: for example, raising funds for treatment overseas, or for increasing public understanding of a condition. This may also have encouraged children to contribute stories to patient information resources.[11]

Finally, post-war theories of child development and their faith in the power of narrative to influence psychic maturation have made *stories* the form of choice for communicating with children. Children were and are encouraged to read stories and act them out, by experts, educators and family care-givers, to widen experience of human nature and to nurture empathy.[12] Twentieth-century children's literature developed well-defined norms, with heroes and villains, crisis and resolution, and happy endings.[13] In the middle decades of the century, comic books became very popular, characterised, as they were, by innovations in art and plot, and catering to different ages of reader as well as to boys and girls as separate audiences. More recently, comics have declined in popularity but children's illustrated paperback books still sell exceptionally well. Cartoons continue to appeal and pictures help to carry younger children through story-lines.[14]

These four traditions – case-studies, prognoses, media and children's stories – have fused in the patient information produced for children with cancer since the mid-1970s. It can be seen that information for children has *not* been written primarily to permit informed choice about treatment; indeed, children are legally disqualified from making such decisions.[15] Rather, information has served to teach readers how to cooperate in ward and treatment room, and to offer them ways of constructing experiences so that these make sense to them and to their loved ones.

## Child patients

The history of child health has been marginal to the history of medicine. But this has been changing since the 1990s. The rise of paediatrics and allied professions in Britain has yet to be thoroughly mapped. Increasing concern for child welfare through the nineteenth and twentieth centuries has, however, been documented.[16] The present paper, which focuses on recent professionalisation around the needs of the sick child, builds on Harry Hendrick's analysis of post-war concerns for the welfare of children in hospitals, and on David Armstrong's historicising of the patient's view. After the foundation of the NHS, the number of children treated in hospitals was expected to rise. In 1948, leading child psychologists and psychoanalysts at the Tavistock Institute, including John Bowlby, Anna Freud and James Robertson, began researching how maternal care affected the development of personality in infancy. Their focus was on the emotional needs of very young children and the likely consequences if children were denied a nurturing home environment. The Ministry of Health consulted the Institute on the likely effects of hospitalisation and separation on children's emotional well-being. The Institute's report suggested that traumatic medical experiences and being deprived of family contact would lead to long-term damage to children's personalities.[17] The Ministry of Health responded by urging hospitals to allow daily visits to child patients by their parents, but few hospitals heeded the call.

In 1956, a committee, under the chairmanship of Sir Harry Platt, was appointed to evaluate how children should be cared for in hospital. Its report in 1959 recommended that hospitals cease admitting children to adult wards, and that children's wards adopt open visiting. But many hospitals were slow to change their practices.[18] Robertson's study of the effects of mother–infant separation on toddlers was filmed for the BBC and broadcast in 1961, along with a series of newspaper articles. This fuelled public calls to reorganise care for sick children in order to protect their psychological well-being. The action group Mother Care for Children in Hospital was founded in the same year to campaign for open visiting and the provision of appropriate recreational facilities within hospitals. The group changed its name in 1963 to The National Association for the Welfare of Children in Hospital (NAWCH). One of its earliest publications was a leaflet for parents to help them prepare children for hospital stays and support them on their home-coming.[19]

By the early 1970s, however, a tide of change was sweeping through children's wards. Other voluntary bodies and a growing number of hospitals had joined NAWCH in producing information materials to prepare children for admission. Save The Children had now been funding play leaders in children's wards in London hospitals for more than a decade and such posts were starting to appear in larger provincial hospitals.[20] Gradually, play specialists began to take a role in paediatric medical and surgical wards, and they helped children to normalise the experience of being ill, in part through creating and distributing patient information resources.

The treatments prescribed for children with cancer kept them in hospital for weeks or months at a time. These patients were some of those whose psychological needs were championed by the charities for child hospital welfare. New charities were formed specifically for this patient group. Malcolm Sargent Cancer Care for Children, founded in 1967, put most of its funds into providing support workers for children's cancer wards, to help families manage and make sense of the experience of treatment.[21] Within a decade it was joined by many others, often started by groups of parents of children with the disease. By the end of the 1970s, children with cancer had secured public attention and support on the grounds that they were peculiarly at risk of psychological damage as a result of their illness experiences.

## Childhood cancer

Before the 1940s, childhood cancer received little attention from the medical profession or the public. Other conditions loomed larger in the popular imagination and paediatricians' consulting offices. In the late 1930s, James Ewing in New York had built up a team of specialists who recorded cancers in children of kinds rarely seen in adults. The findings were published in 1940. Physicians interested in the cancer problem visited Ewing's children's ward at Memorial Hospital, New York.[22] These included British radiotherapist, I. G. Williams.[23] Williams surveyed childhood cancers in Britain, and published his results in the *British Journal of Radiology* in 1946, stimulating interest among pathologists and paediatricians.[24] At this point, survival rates reported by British surgeons were around 10 per cent, but Memorial and other leading paediatric hospitals in America were recording much higher figures. These latter institutions were associated with treatment that integrated surgery, radiotherapy and supportive care.

Leukaemia at this time was invariably and rapidly fatal. Sydney Farber of the Children's Hospital in Boston developed chemotherapeutic approaches for cancers beyond the reach of surgery and radiotherapy, and by the end of the 1940s he was reporting increased survival for leukaemic children and patients whose tumours had metastasised. Pathologists and paediatricians from British teaching hospitals sought temporary fellowships to study in Boston and other leading cancer research centres in America. They organised small-scale leukaemia and tumour clinics on their return to Britain. By the end of the 1950s there were well-established specialist clinics in London, Newcastle and Manchester.[25]

The availability of antibiotics to treat infections led to dramatic reductions in infant deaths in the 1950s, elevating cancer to the second largest cause of death for children over the age of one. In the 1950s experiments with anti-cancer drugs showed greatest success against cancers peculiar to childhood. Childhood cancer thus became a hot topic for research, leading to the formation of a new discipline of 'paediatric oncology', new patterns of cooperation in and between hospitals in coordinating treatment and running clinical trials of innovative therapies, and radically new experiences for children diagnosed with a malignancy. A typical patient in the early 1950s stayed in an adult ward under the care of a surgeon. By the late 1970s, most children with cancer were treated together in dedicated wards in specialised centres in which teams of physicians and surgeons delivered high technology intensive multi-modal therapy.

Cancer 'cures' tend to carry precisely those quote marks in medical writings since the return of the disease cannot be ruled out, and even successful treatment can do lasting damage to health. In the case of children, paediatricians' commitment to facilitating normal psychosocial development multiplied difficulties in talking about 'cured' patients. Children were seen to be psychologically scarred by their experience as cancer patients: treatment disrupted family life, and many children suffered lasting side-effects that compromised their skills and self-confidence. By the mid-1970s, approximately 50 per cent of children were surviving cancer. Ensuring that their personalities emerged undamaged became increasingly important. Conferences were organised on the problems of the cured, and textbook chapters were added on minimising the cost of prolonged treatment and uncertainty about the future. Psychologists were commissioned to survey adjustment among the cured and to identify forms of preventive help that might be offered to future patients.[26]

The proliferation of patient information for children with cancer was part of this reaction to rapidly changing cure rates. As more children survived, charitable funds and psychologists' efforts were diverted away from producing books helping parents prepare for grief, and towards distributing materials to assist families to cope with uncertain survival. 'Psycho-oncology', the study of how patients and their families fared psychologically and socially after a diagnosis of cancer, emerged as a new sub-specialism at the same time and in the same places as information resources for child patients.[27] Discussions in medical journals about whether it was best to tell children the truth or to lie about prognosis terminated in the early 1980s as anthropologists and sociologists insisted that children often already knew about their diagnosis and chances of survival.[28] Over the same period, increasing numbers of cheerful little booklets were being handed out to the newly diagnosed, explaining cancer, its treatment and outcomes.

## Cartoon booklets and their readers

We have seen that booklets were produced for parents with children needing hospital admission from the 1960s onwards, and that by the early 1970s the supposedly unique needs of children with cancer were being defined. The new interest in children's cancers and in the psychological well-being of child patients came together in the mid-1970s in the form of psychologically informed efforts designed to help children cope with cancer and its treatment. Over the course of the following decade, a standard and largely didactic approach developed, in which a featured patient or group of patients moved from diagnosis through treatment to the resumption of normal life.

Significant changes of genre have occurred over the past 40 years of patient information. Materials include those provided by British charities and by specialist hospital units, a small number of American resources distributed in Britain, patient and donor magazines, online materials and privately circulated photocopies of patients' comics.[29] Until the late 1990s, most booklets were written by psychologists, and a minority by graduate paediatric nurses. However, resources circulated since then have been authored by a more diverse group of writers. The most common format has remained constant: a story, following a typical patient from diagnosis through ward and out-patient treatment to follow-up. These stories attempt to explain cancer to children at three different levels: intra-patient, exploring the relation between the child and his or her cancer cells; inter-patient, showing children sharing their experiences to relieve fear and normalise experience; and patient–doctor interaction, encouraging patients to think of themselves as, or to be, active participants.

The level that predominated, the amount of information offered, and the style of presentation varied according to age of intended audience. Material produced for children under seven concentrated on the first level. Young children were known to fear pain and to tire of books quickly; authors therefore focused on helping readers through each needle and nasty taste. The second level came to the fore in resources for older children and adolescents. These patients were deemed more susceptible to fear of death, and more dependent on circles of friendship. The third level appeared in information aimed at an exclusively adolescent audience, a cohort keen to practise autonomy yet legally not permitted to do so. Information for these patients encouraged assent to treatment and stressed mutual respect between medical team and adolescent. Materials for the youngest children have not changed over the period in question, but those for the older groups have: this shift has been brought about by changes in authorship since the late 1990s.

Individual treatment centres, national professional bodies and childhood cancer charities started freely distributing booklets for children in the mid-1970s. Initially, British clinicians relied on imports of *You and Leukemia*, by the psychologist, Lynn Baker, and published by the National Cancer Institute (NCI) in 1975.[30] This was the first information booklet produced for child cancer patients. A few treatment centres produced short leaflets or colouring sheets explaining treatments and the disease process for their patients in the 1970s, but the first British *stories* did not become available until the mid-1980s. These were produced for and distributed by the charity Sargent, and by Manchester's treatment centre.[31] They were modelled on *You and Leukemia*. New strategies for explaining cancer and its treatments emerged in the late 1990s as the needs of adolescent patients received more attention. Patient information booklets for children have now proliferated: CancerBACUP's 2003 survey of materials for children and adolescents with cancer located over 200 different texts in use across Britain.[32]

The oldest booklets were aimed at children below the age of ten, using simple words and portraying cells as cartoon characters.[33] For example, the series of books produced by Manchester's paediatric oncology services in the mid-1980s used the 'Mr Men' characters of Roger Hargreaves to represent patients and cells: Mr Clever served as an image for white blood cells, smart enough to know how to deal with germs, and Mr Bump represented platelets, an essential blood component. In these materials, each type of cell had a job title that would be familiar to young children: red blood cells as lorry drivers, for example, and platelets as policemen making sure blood stayed within its proper channels.[34]

From Baker's book onwards, the body was presented as a complex system of factories – groups of cells working together – with a distribution network of blood and lymph carrying products to where they were needed.[35] Cancer was portrayed as the result of cells not knowing what they should be doing – leukaemia as a cluster of symptoms caused by the body making too many immature white blood cells, and tumours as clumps of confused cells sticking together for company.[36] Younger children were reassured that cancer cells were not malevolent but rather that their bodies were making the wrong kinds of cell by accident. Cancer cells

were described as merely badly behaved, growing too quickly, crowd-
ing out cells from which they differed.[37] This approach supposedly discouraged 'splitting-off', the notion that a child who feels badly let down
by his or her body will develop profound self-hatred leading to long-term
mental health problems. Psychologists sought to prevent developmental
damage in children by presenting cancer as belonging to 'self', not 'other'.

Such a characterisation of cancer was much less sinister than that
depicted in stories about cancer in adults, where the condition was typically presented as an invading or alien life-force, against which the victim
must fight. Aggressive imagery did, however, appear in material written
for and by older children and adolescents. These patient groups were not
provided with information by charities and hospitals until the 1990s,
although they had long been seen as having information needs different
from those of younger children.[38] Since the Platt Report of 1959, successive government bodies had recommended that children over the age
of 13 or 14 be treated in dedicated units away from very young children.
Since the mid-1970s, conferences on the psychosocial needs of the cured
had stressed that adolescents had additional concerns about the effects of
cancer and treatment on their independence, sexual potency and future
employability – significant components of developing identity – and that
they required reassurance that they could still exercise control and a degree
of autonomy. In the 1990s, psychologists sought and secured funding for
specialised information resources for these patients, at a time when professional structures throughout the health service finally began to provide
dedicated facilities for adolescents.[39]

Psychological support workers, employed by children's cancer charities
to work in treatment centres, asserted that for older children and adolescents the source of such reassurance was of great importance: information
from within the medical system was suspect.[40] Information from consultants and nurses was 'contaminated' because it came from a world that
demanded renunciation of self-direction in the service of compliance
with treatment protocols. Thus patient-authored materials were seen as
the only route to reach older paediatric patients. A few older children and
teenagers produced personal accounts of cancer in short stories or comic
strips that were copied for use within their hospitals. By the end of the
decade, hospitals and charities were investing in making these items more
widely available. In these latter materials, depression, jealousy, fear of
death, disfigurement and constipation secured greater page space.

The introduction of children into the production process changed the
ways in which information was presented. Fictional patient experiences
were rejected in favour of true stories. Some organisations also started

consulting younger children about their information needs, although the cartoon booklet has remained the typical form in which information is delivered for this age-group. Projects such as the Leukaemia Research Fund's *Jack's Diary*, published in 2000, and the Brain and Spine Foundation's *Headstrong* Project, launched in 2004, featured actual children, not tidied stories or composites, all at the insistence of the children involved.[41] In the latter, photographs were requested rather than cartoons; in the former, each page included a flap, under which there was a photo of Jack or of his handwritten copy to prove that the story had not been made up.

In order to reach older children and adolescents, hospitals and charities have made use of new techniques for sharing patients' experiences: CD ROMs, information folders and web-based stories and games. The World Wide Web has proved particularly successful as an information medium. Families have used the web to keep friends informed about their children's progress, and to locate and be located by others with similar experiences. The facility for connecting sites in endless chains of hyperlinks enabled collections of stories to be built up.[42] United Kingdom Children's Cancer Study Group (UKCCSG), the professional body for paediatric oncologists in Britain, and several hospitals and charities, host websites that

*Figure 5.1* The first panel of the original comic, *Captain Chemo*, 1999. The Royal Marsden NHS Foundation Trust. Reprinted with permission.

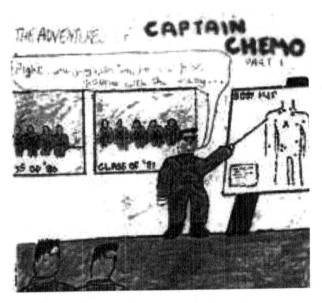

offer links to stories from their own patients and web logs of families elsewhere.

Psychologists have argued that writing stories helps patients to make sense of experience by putting it into words and giving it broader and public meaning.[43] Contributors to Great Ormond Street's website 'Children-first' explained that they shared their tales to raise awareness among other patients and the public at large.[44] These networks of cancer stories inform patients and increase the visibility of childhood cancer. In the process of chaining tales together, editorial control falls back into the hands of the technical experts in childhood cancer, as they select stories most likely to be of use in representing these conditions in helpful ways.

Older children and adolescents have radically changed the way in which chemotherapy is represented to new patients. Booklets for younger children presented chemotherapy as restoring bodies to healthy balance, but adolescents have envisaged drugs as weapons to eradicate an enemy life-force cell by cell. Full use has been made of new media to distribute this alternative narrative.

## Representations of chemotherapy: Captain Chemo and Mr Wiggly

Since the late 1970s, chemotherapy has been a part of the treatment programme for almost all children with cancer. Verbal and printed information likened chemotherapy to weed-killer. The analogy originated in Memorial Hospital in New York, when medics described the chemotherapy being developed for child patients through the late 1940s as working within the body against cancer in the same way that weed-killer worked in farms to increase the health of a crop.[45] Booklets of the 1980s and 1990s developed this analogy: Simon, a ten-year-old narrator, told his audience that 'The Doctor and Nurse ... said that the body is a bit like a garden. In the garden there are lots of types of flowers ... But in *my* body I also had cancer cells which were like weeds.' The illustrations showed a spring garden strangled by weeds at diagnosis, and returned to beauty through chemotherapy.

We have seen that information for younger children painted cancer as a muddle not an invasion, but in more recent materials, for and by older children, cancer has resembled adult representations: cancer as an 'other' to attack. The most popular information resource for explaining chemotherapy to older children is the 'Captain Chemo' website. Captain Chemo first hit the alternative comics scene in 1999. His creator drew a comic strip about the Captain in 1994, while receiving treatment at the

Royal Marsden, since he was unable to find materials specifically for teenagers. His family gave the copyright to the hospital. Other Captain Chemos have been created, but by far the most famous, if website hits are anything to go by, is this hero, accounting for half of the 1,800 sites located by Google and featuring a Captain.[46] Paper copies of the comic were sent to cancer centres, then funds were raised to develop a series of interactive episodes on the hospital's website where children take part in the Captain's missions by answering questions, steering the chemo craft, and shooting tumours.[47] In the translation, Captain Chemo was transformed by a graphics team from a regular-looking military man into a stereotypical hunk. The mediation of chemotherapy by a lay expert was re-mediated to conform more closely to children's assumed expectations of (male) superheroes.

*Figure 5.2*   The website version of *Captain Chemo*, 2001. Royal Marsden NHS Foundation Trust. Reprinted with permission.

The website is pitched at the eight to 14-year-old age range, a prime target for computer games that invite or order that the player shoot at anything 'not self'. Other web-based computer games by and for cancer patients that simulate active killing of cancer cells through firing chemo weapons have since been developed.[48] The resumption, or illusion, of control over treatment choices and efficacy has enormous appeal for patients moving towards adolescence.

*Figure 5.3*    Jack's in-dwelling catheter was represented to him, and in the book based on his experiences, as a friendly creature: 'So that you don't have to have too many needles the doctors put you to sleep and put a Hickman Line into your chest. ... We called him Mr Wiggly'. *Jack's Diary*. The Leukaemia Research Fund. Reprinted with permission.

The success of Captain Chemo and his comrades may lie ultimately in the fact that their authors cannot be doubted to be 'telling it like it is'. There is no suspicion that the story of living through cancer treatments has been doctored to make it seem more palatable. Captain Chemo in original comic form and on the website deals openly with unpleasant side-effects such as constipation and weight gain. What is missing, however, is patient experience itself. Readers or players are invited to watch what is happening inside the patient through the monitors being watched by the Captain and his Cadets, and to control it, not to experience it from the inside.[49] The abstraction from personal experience and assumption of an alternative external and powerful point of view offers a double relief to children undergoing cancer therapy.

Since the 1970s, in-dwelling catheters that have an external connection to the blood supply have been inserted in most children with cancer to make the taking of blood and giving of chemotherapy and transfusions less distressing. This prosthesis is a heavy burden for a young child to endure. It requires surgery to be inserted, frequent maintenance, and marked changes in routines such as bathing, exercising and self-fondling. In order to make the line less scary, it has been introduced with a pet name, typically 'Wiggly' or 'Mr Wiggly'. Colouring books and stories about this character were generated locally by play therapists for distribution to new patients.[50]

Wiggly was and is a snake, caterpillar or worm, with a smiley face and a head into which needles can be stuck without pain, delivering nice 'drinks' of chemotherapy. Patients were told Wiggly was happy to visit for a while to help by accepting injections on their behalf but could only stay if kept dry and if not pulled. Wiggly proved less successful as a mediating trope for older children. Play-leaders report that the representation turned off children who considered themselves too mature for such an infantile character, or too wise to accept that needles in the head would hurt less than needles in the arm. Rather, older children have represented this technology more accurately. Jack of *Jack's Diary* wrote about Wiggly in his own personal story. But the 12-year-old author of *Chemo Girl* developed a more realistic characterisation, in the form of a super-hero partner for Chemo Girl, Superdog 'Brouvi'. The dog's jaws are in fact the two ends of a double-lumen catheter.[51]

## Conclusion

The treatment of cancer in children over the last 30 years has been high-tech, high-risk and highly emotive. Multi-modal technological fixes developed quite suddenly, creating a new breed of doctor and a new kind of patient who had to find ways of communicating with one another. Those patients whose needs were the most pertinent to psychologists have shaped how child patients have been addressed. Initially, the focus was on the youngest children, who were seen to face psychological damage if treatment were allowed to disrupt the routines of family life. Booklets were produced that attempted to normalise cancer and to help children cope with the stresses of treatment. Over the period as a whole, however, the cohort whose needs have shaped information production changed from the youngest to the oldest. The more recent innovations in patient information have originated in efforts to serve the psychological needs of adolescents.

Psychologists, hoping to help children adjust to prolonged medical attention and manage uncertainty about the future, penned booklets intended to explain cancer and patienthood. This didactic approach was seen to fail for older children, who declined to take their booklets home and drew their own stories. Medical charities and doctors' professional bodies used patient-authored materials to engage older children and adolescents. Psychologists and the patients they represented stressed the value of producing information. Contributing stories was perceived as a way in which older children and teenagers could resist the passivity that patienthood seemed to imply, preserving normal psychological development and progressing towards autonomous adulthood.[52] The chance to 'tell it like it is' came to be seen as medically beneficial.

Children with cancer found that their special needs gave them a way publicly to record their experiences of illness. The history of child health has tended to approach its subjects as a passive group of medicalised patients, subject to others' definitions of experience and meaning.[53] But, as this article shows, in recent times, and in some places, sick children have acquired limited agency in the historical process, and have begun to shape how they are seen within the medical networks through which they move.

## Acknowledgements

This work was funded by the Wellcome Trust as part of the Programme Grant 'Constructing Cancers, 1945–2000'. I am grateful to my colleagues on this project, Elizabeth Toon, Carsten Timmermann, Helen Valier and John Pickstone, and two anonymous reviewers, for their comments on earlier versions of this article. I am indebted to Ayesha Nathoo for her insights on the role of the human interest story in the media in the late twentieth century. I would like to thank the play specialists in the paediatric oncology centres across Britain and in Eire, and the information officers at CancerBACUP, Sargent Cancer Care for Children, the Leukaemia Research Fund and the Royal Marsden Hospital, for all their assistance in locating information materials.

## Notes

* This chapter was originally published in *Social History of Medicine*, 19, 2006, 501–19. It has been slightly revised for this book.
1 On the history of how children's needs have been understood in Britain, see Woodhead 1997.

2   A bibliography is provided of the most frequently distributed titles, as reported by patient information providers and users.

3   Patient classification in cancer registries and clinical trials frequently place the top end of childhood at 15. For this reason, the phrase 'children and young people' has come into use since the mid-1990s, although this has already begun to date.

4   In July 2004, 'Omega Boy versus Doctor Diabetes' was launched by two sons of a comic producer. One explained 'When I was in the hospital, I didn't have anything to read (about diabetes) that was for the kids so I made up a comic book because I love comics', Patty, S. http://www.silverbulletcomicbooks.com/news/109121662115031.htm, 30 July 2004 (last accessed 23 August 2011).

5   The classic starting-points for an exploration of the secondary literature on the meaning of cancer are Sontag 1979 and Patterson 1987. Neither consider that cancer might be represented to and by children differently from the manner in which it is seen by adults.

6   This is beginning to change. Women's magazines occasionally publish pieces on breast cancer that present information on screening, treatment and outcome through the medium of a story of another patient's progress. These pieces are written not as 'true life' stories but as medical information. See, for example, Marchetto 2005, pp. 260–5.

7   Hunter 1991. On the value that telling one's story can have for a patient, see Hawkins 1993 and Frank 1995. Armstrong 1984 offers a history of the nature and purpose of case-taking in twentieth-century medical practice.

8   Published memoirs by British parents were rare before the 1980s; the best known is Smith 1964. Interviews with eight paediatric oncologists, conducted in London and Manchester, 2004 and 2005.

9   Information sheets were produced by paediatric oncologists from the 1960s; interviews with eight paediatric oncologists, conducted in London and Manchester, 2004 and 2005. In 1979, the United Kingdom Children's Cancer Study Group issued a booklet for national distribution. See minutes of meeting held in Birmingham, 8 June 1979.

10  Classic works on illness memoirs include Kleinman 1988 and Brody 1987. On the origins of the human-interest story in the media, see Thompson 2000. See Hansen 1998 for an account of the use and manufacture of celebrity status in medical reporting.

11  The progress of one sick British child, whose family sold up to move to Texas to take part in a clinical trial, was followed by the BBC: http://news.bbc.co.uk/1/hi/england/kent/3141516.stm, 26 September 2003, explains the family's decision; the last story is http://news.bbc.co.uk/1/hi/england/kent/3934721.stm, 29 July 2004 (sites last accessed 23 August 2011).

12  For a scholarly and personal reflection on the impact on an adult of reading in childhood, see Spufford 2002.

13  The same structures have dominated patient information materials since mass distribution began in the 1970s. As one survivor wrote in her account of her treatment for osteosarcoma, 'Most stories have a happy ending and this was no exception'. 'Contact: A helping hand for families of children and young people with cancer', United Kingdom Children's Cancer Study Group (UKCCSG): http://www.cclg.org.uk/contact/PDFs/Contact18.pdf, p. 5. (Last accessed 23 August 2011).

14 Corr 2002 lists 168 books on death dating back to 1949, subdivided by age of reader. The books for children under the age of ten are all richly illustrated.

15 When a patient consents to medical or surgical treatment, she states that the actions performed on her person do not constitute assault. This right to autonomous control of one's person is consequent upon mental competence. British courts have debated when children attain this competence and how it should be proved since the mid-1980s. See Alderson 1993.

16 These include Cooter (ed.) 1992, Gijswijt-Hofstra and Marland (eds) 2003, Hendrick 1994, 1997, and Steedman 1990.

17 Robertson 1958.

18 Platt 1959.

19 Barnes 1999.

20 Some voluntary hospitals employed women to coordinate play in wards from the 1950s. The first designated play-leaders outside London were hired in 1970. Interview with Pamela Barnes, trustee of Action for Sick Children, 2004.

21 This charity later changed its name to 'Sargent Cancer Care for Children' and in 2005 merged with CLIC (Cancer and Leukaemia in Childhood).

22 Dargeon (ed.) 1940.

23 Taylor 1990.

24 Williams 1946. Interviews with eight paediatric oncologists, conducted in London and Manchester, 2004 and 2005.

25 Interviews with eight paediatric oncologists, conducted in London and Manchester, 2004 and 2005.

26 Most of these meetings and studies were American, since there were more survivors and more state funding for research in that country. But their findings were noted in British paediatric oncology units. An annotated bibliography of studies on the 1970s can be found in Koocher and O'Malley (eds) 1981.

27 For the history of research into the psychological impact of childhood cancer, see Eiser 2004.

28 The argument can be traced in, for example, *The Lancet* through the 1960s and 1970s. Bluebond-Langer 1978 has been credited with offering the most convincing evidence of children's knowledge of their illness.

29 The author surveyed the 22 paediatric oncology centres in Britain and Eire, contacting play-leaders by letter in April 2004. The response rate was 77 per cent.

30 Baker *et al.* 1975.

31 Interviews with eight paediatric oncologists, conducted in London and Manchester, 2004 and 2005.

32 Survey conducted by CancerBACUP, 2003 to 2004.

33 Personification is not a new strategy in educating children about health. Feldberg describes a flurry of information booklets about tuberculosis for children in the 1910s, for circulation within sanatoria; pamphlets starred disease agents 'Tommy Tubercle' and 'Huber the Tuber'. Feldberg 1995.

34 Both examples come from Baker *et al.* 1975, but were repeated in all the major British booklets of the 1980s and 1990s.

35 A British innovation was to add roundabouts to the distribution network metaphor: Colinshaw [late 1980s].

36 Dales and Pesterfield 1996 and Colinshaw [late 1980s].

37   Baker *et al.* 1975 pictured spiky cancer cells stuck together on one corner of
     the page, while the better adjusted (and smooth) cells stayed away.

38   The Royal Victoria Infirmary, Newcastle organised creative writing groups
     for adolescent cancer patients and their parents in the 1990s, but the stories
     produced were not distributed outside the region.

39   See Barnes 1999 for a bibliography of the many reports of the 1990s on the
     distinct medical and psychological needs of adolescents.

40   Sargent's social workers offered counselling from the early 1970s. CLIC funded
     play-leaders from 1976. The Teenage Cancer Trust began to pay for activity
     leaders in its new adolescent units from 1995. Since the 1970s, many smaller
     children's charities have sent clowns or toys into wards, or paid for parties.
     This is in addition to psychological support provided by the NHS.

41   The Brain and Spine Foundation launched *Headstrong* on 18 March 2004 with
     a ring-bound folder for younger children, a comic for older children, a CD,
     and a website http://www.headstrongkids.org.uk/ (last accessed 23 August
     2011). This has been described as the most expensive piece of patient informa-
     tion produced thus far. Much time and consultation were required before
     ethics committees gave clearance to the charity to involve ill children in the
     project.

42   Children's Cancer Web: http://www.cancerindex.org/ccw/. This is an inde-
     pendent non-profit site, established in 1996 to provide a directory of child-
     hood cancer resources. There are in total 78 home pages, classified by cancer
     type (site last accessed 23 August 2011).

43   Bolton 1999. This author works with adolescent cancer patients and uses
     creative writing, often based around the experience of illness.

44   Great Ormond Street hosts a website on which child patients discuss the
     pros and cons of treatment options: http://www.childrenfirst.nhs.uk/ (last
     accessed 23 August 2011).

45   In *Hygeia* in December 1946, William Dameshek, a haematologist at Memorial
     Hospital, New York, described a person with acute leukaemia as being like
     a healthy garden overtaken by 'an overgrowth, like a weed, of a special type of
     white cell'. Krueger 2003, pp. 39–40.

46   A Google worldwide search on 31 January 2005 for 'Captain Chemo' yielded
     1,810 hits; 38 of the first 50 were for this version. With the additional search
     term 'Marsden', 859 pages were identified. Since this chapter was first pub-
     lished as an article in *Social History of Medicine* in 2006, at least one other
     patient-created Captain Chemo has entered the frame, this one a character in
     an adult's memoir: Vickers 2009.

47   In July 1999, the Royal Marsden printed and distributed 5000 copies of the
     comic. The first interactive website was ready in April 2000. For the current
     version, see http://www.royalmarsden.org/captchemo/ (last accessed 23 August
     2011).

48   Ben Duskin, aged nine, asked the Make A Wish charity in San Francisco to pro-
     duce a video game to explain cancer to young children. LucasArts video games
     created the game exactly as Ben wanted. http://news.bbc.co.uk/1/hi/ technology/
     3839091.stm, 28 June 2004. The game is available at http://www.sfwish.org/
     site/pp.asp?c=bdJLITMAE&b=81924 (sites last accessed 23 August 2011).

49   There is a risk that players may fare badly and be unable to hit the tumour
     cells fast enough to prevent recurrence, which could make a child feel more

anxious. No advice is offered to adult helpers on how to avoid or deal with this eventuality.

50  In Our Lady's Hospital for Sick Children, Dublin, in-dwelling catheters are known as 'Freddie', named after one of the consultants who places them. Ann Coyle, the hospital's play specialist, reported that children of all ages are encouraged to use this name. Personal correspondence, 14 October 2004. Ann Coyle also produced a booklet to help children prepare for the insertion of a Hickman catheter, *Gillian Gets her Freddie*.

51  Richmond 1997. Unfortunately, due to copyright issues we cannot reproduce the picture of Chemo Girl included in the original version of this chapter in *Social History of Medicine*.

52  Enforced patient passivity has long been recognised as a major cause of subsequent psychological adjustment problems. See, for example, Zubrod 1975, p. 267.

53  Hendrick is correct to stress that the social construction of childhood has been a process governed by the professional middle classes. Hendrick 1994, p. 19. But it can be argued that post-war psychological attention to the needs and potentials of children and adolescents has led to a present in which subjects have become genuinely autonomous historical agents.

## Works Cited

### British resources

Barton V. late 1990s or 2000, *Joe Has Leukaemia*, illustrated T. Harris and R. Jago, Sargent Cancer Care for Children.

Barton V. late 1990s or 2000, *Mary Has a Brain Tumour*, illustrated T. Harris and R. Jago, Sargent Cancer Care for Children.

Barton V. late 1990s or 2000, *Lucy Has a Tumour*, illustrated T. Harris and R. Jago, Sargent Cancer Care for Children.

Blee S. 1993, *Getting Rid of Groggle: I'm Fighting Cancer*. Jennifer's Story, Luton: Luton Borough Council.

Brazier L., Trapp A. and Yates N. 1990s, *Simon Has Cancer*, illustrated P. Yates, Newcastle: Royal Victoria Infirmary.

Colinshaw S. late 1980s, *Jenny Has a Tumour*, illustrated C. Gale, Nottingham: Malcolm Sargent Cancer Fund for Children.

Coughlan S. and Cuddy C. 1995, *A Child's Guide to Radiotherapy*, Glaxo.

Dales J. and Pesterfield C. 1996, *Leukaemia Made Simple*, Cambridge: Addenbrooke's Hospital.

De Garis B. ['aged 13$\frac{1}{2}$'] 1999, *The Adventures of Captain Chemo and Chemo Command*, Surrey: Royal Marsden NHS Trust.

Dempsey S. 2003, *My Brain Tumour Adventures: The Story of a Little Boy Coping with a Brain Tumour*, illustrated G. Collins, London and Philadelphia: Jessica Kingsley.

Hague A. 1985, *Leukaemia*, Fareham: Lederle Laboratories.

Hague A. and Silkstone J. 1989, *When You Have a Tumour*, Manchester: Inter-City Publications.

Hubbard D. 2002, *Jack's Diary*, illustrated N. Humphreys, Leukaemia Research Fund.

Lambdin J. [aged 10] 1996/7, *Jodi's Story*, Woking: Design Alliance.

Nicholson A. and Thompson J. 1980s, *Jenny Has Leukaemia*, illustrated N. Routledge, Malcolm Sargent Cancer Fund for Children.

Palmer S. 1994, *Fighting the Big 'C': A Guide for Young People and Their Families to Understand Cancer and its Treatment*, sponsored by Cancer and Leukaemia in Childhood.

Pearman K. E. 1998, *Jasper the Rabbit and Polly the Giraffe: A Story to Help Children Undergoing Chemotherapy Manage Their Nausea and Vomiting*, illustrated C. Hamilton, sponsored by GlaxoSmithKline.

Ronnie P. [Purple] 2001, *A Young Person's Guide to Lymphoma*, The Lymphoma Association.

Silkstone J. and Hague A. 1985, *When Your Sister or Brother Has Leukaemia*, Lederle Laboratories.

Silkstone J. and Hague A. 1989, *When Your Brother or Sister Has a Tumour*, Regional Oncology Support Service Manchester and David Bull Laboratories.

Vickers P. 2009, *Fighting Captain Chemo: A Story of Love, Loss and Lymphoma*, Ferrybridge: Pen2Print.

*Wiggly's Story* (photocopied colouring book: no author, no publisher, no date).

## North American resources distributed in Britain

Baker L. S. in collaboration with Roland C. G. and Gilchrist G. M. 1975, *You and Leukemia: A Day at a Time. For Children with Leukemia (and Other People)*, Bethesda, Maryland: National Cancer Institute.

Baker L. S. 1978 [1988 rev.], *You and Leukemia*, Philadelphia and London: Saunders.

Cranston L., LeBlanc C., Restivo M. and Barr R. 2001, *You and Your Cancer: A Child's Guide*, Sponsored by Ronald McDonald House Charities, Hamilton Ontario and London.

Crowe K. 2000/1, 'Me and My Marrow: A Kid's Guide to Bone Marrow Transplants', illustrated N. Bendell, www.fujisawa.com (last accessed 31 January 2005).

Krisher T. 1992, *Kathy's Hats: A Story of Hope*, illustrated N. B. Westcott, Morton Grove Illinois: Albert Whitman.

Marston S. 2000, *Sarah: A Six-Year-Old Who is Unafraid of Cancer*, Integrity and the National Childhood Cancer Foundation in North America.

Richmond C. 1997, *Chemo Girl: Saving the World One Treatment at a Time*, Sudbury Massachusetts: Jones and Bartlett Publishing.

Romar P. 1998, *Sam Fights Back*, Leukemia Society of America.

Rosenberg D. 1996, *The Talking Lady Presents: Having a Brain Tumour*, illustrated T. Mortenson, Toronto: The Talking Lady Press.

Westcott P. 2000, *Living With Leukemia*, Austin, Texas: Raintree Steck-Vaughn.

Wurtz J. 2001, *Wilm's Tumor: What Now?*, Kidney Cancer Association.

## Resources for relatives

Colinshaw S. 1993, *When My Little Sister Died*, illustrated C. Gale, Nottingham: Merlin Publishing.

Stokes J. A. 2000, *The Silent C: Straight Talking about Cancer*, illustrated P. Bailey, Winston's Wish in association with Macmillan Cancer Relief, Gloucester Royal Hospital.

## Secondary sources

Alderson P. 1993, *Children's Consent to Surgery*, Buckingham: Open University Press.

Armstrong D. 1984, 'The Patient's View', *Social Science and Medicine*, 18, 737–44.

Barnes P. 1999, *Royal Manchester Children's Hospital, Pendlebury, 1929–1999*, Manchester: Churnet Valley Books.

Bluebond-Langer M. 1978, *The Private Worlds of Dying Children*, Princeton: Princeton University Press.

Bolton G. 1999, *The Therapeutic Potential of Creative Writing: Writing Myself*, London: Jessica Kingsley.

Brody H. 1987, *Stories of Sickness*, New Haven and London: Yale University Press.

Cooter R. (ed.) 1992, *In the Name of the Child: Health and Welfare, 1880–1940*, London: Routledge.

Corr C. A. 2002, 'An Annotated Bibliography of Death-Related Books for Children and Adolescents', *Literature and Medicine*, 21, 147–74.

Dargeon H. (ed.) 1940, *Cancer in Childhood*, St Louis: Mosby.

Eiser C. 2004, *Cancer in Children: Quality of Life*, London: Lawrence Erlbaum.

Feldberg G. D. 1995, *Disease and Class: Tuberculosis and the Shaping of North American Society*, New Brunswick, NJ: Rutgers University Press.

Frank A. W. 1995, *The Wounded Storyteller: Body, Illness, and Ethics*, Chicago: University of Chicago Press.

Gijswijt-Hofstra M. and Marland H. (eds) 2003, *Cultures of Child Health in Britain and the Netherlands in the Twentieth Century*, Amsterdam: Rodopi.

Hansen B. 1998, 'America's First Medical Breakthrough: How Popular Excitement about a French Rabies Cure in 1885 Raised New Expectations for Medical Progress', *American Historical Review*, 103, 373–418.

Hawkins A. H. 1993, *Reconstructing Illness: Studies in Pathography*, West Lafayette: Purdue University Press.

Hendrick H. 1994, *Child Welfare: England, 1872–1989*, London: Routledge.

Hendrick H. 1997, *Children, Childhood and English Society, 1880–1990*, Cambridge: Cambridge University Press.

Hunter K. M. 1991, *Doctors' Stories: The Narrative Structure of Medical Knowledge*, Princeton: Princeton University Press.

Kleinman A. 1988, *The Illness Narratives: Suffering, Healing and the Human Condition*, New York: Basic Books.

Koocher G. and O'Malley J. (eds) 1981, *The Damocles Syndrome: Psychosocial Consequences of Surviving Childhood Cancer*, London: McGraw-Hill.

Krueger G. 2003, '"A Cure is Near": Children, Families, and Cancer in America, 1945–1980', unpublished D. Phil. thesis, Yale University.

Marchetto M. A. 2005, 'Cancer Vixen', *Glamour*, American edition, May, 260–5.

Patterson J. T. 1987, *The Dread Disease: Cancer and Modern American Culture*, Cambridge MA and London: Harvard University Press.

Platt Sir H. 1959, *The Welfare of Children in Hospital: Report of the Platt Committee*, London: HMSO.

Robertson J. 1958, *Young Children in Hospital*, London: Tavistock Publications.

Smith E. 1964, *To the Bitter End*, London: Abelard-Schuman.

Sontag S. 1979, *Illness as Metaphor*, London: Allen Lane.

Spufford F. 2002, *The Child That Books Built*, London: Faber and Faber.

Steedman C. 1990, *Childhood, Culture and Class in Britain: Margaret McMillan 1860–1931*, London: Virago Press.

Taylor G. 1990, *Pioneers of Pediatric Oncology*, Houston: University of Houston M. D. Anderson Cancer Center.

Thompson J. 2000, *Political Scandal: Power and Visibility in the Media Age*, Cambridge: Polity Press.

Williams I. G. 1946, 'Cancer in Childhood. Abridged from a Thesis accepted by the Fellowship Board of the Faculty of Radiologists', *British Journal of Radiology*, 19, 182–97.

Woodhead M. 1997, 'Psychology and the Cultural Construction of Children's Needs', in A. James and A. Prout (eds) *Constructing and Reconstructing Childhood: Contemporary Issues in the Sociology of Childhood*, London: Falmer, 63–84.

Zubrod G. 1975, 'Successes in Cancer Treatment', *Cancer*, 36, 267–70.

# Part II
# Pathways

# 6

# Knife, Rays and Women: Controversies about the Uses of Surgery versus Radiotherapy in the Treatment of Female Cancers in France and in the US, 1920–1960

*Ilana Löwy*

## Uncertain diagnoses and unruly patients

In 1960, François Baclesse and two of his colleagues at the Curie Institute published a paper entitled 'Should one be authorized to do a simple tumourectomy followed by radiotherapy in a case of a mammary tumour'. Baclesse, the head of the Curie's radiotherapy department, argued that the ablation of a malignant tumour of the breast followed by radiotherapy could be as effective as the standard therapy for breast malignancies, the radical mastectomy.[1] Baclesse's conclusion was grounded in more than 20 years worth of observations, of 100 women with stage I and II breast cancer who had been treated with surgical removal of the tumour alone followed by radiotherapy.[2] Half of the patients remained symptom free for ten years or more, and 17 additional women did not show any signs of breast cancer for at least five years after treatment. The women treated with this conservative approach had slightly higher rates of local recurrence than those who underwent radical surgery, but the five years survival rate was similar in both groups.[3] Tumourectomy (the surgical excision of the tumour) with radiation was less effective than mastectomy in young women (under 35), and those whose tumours were of high histological grade. However, women whose cancers had spread to axillary lymph nodes but not to more distant sites fared relatively well: 55 per cent of the patients with stage II breast tumours (26 out of 47) were stabilised with tumourectomy and radiation alone, as compared to 71 per cent (38 out of 53) of those with stage I breast tumours (small, localised growths without lymph node involvement). Baclesse and his colleagues counted women who had local

103

recurrences as treatment failures, but reminded their readers that these women could still be successfully treated with radical surgery. Finally, they concluded prudently that:

> We are aware of all the objections that may be raised against this method of tumourectomy followed by roentgen therapy, [which is] so violently, so radically opposed to all the classical methods of treatment of breast cancer [...]. We do not intend to propose the generalisation of this method, that remains, in our opinion, an exceptional approach.[4]

The presentation of tumourectomy with radiotherapy as exceptional treatment for breast cancer was modulated, however, by Baclesse's proposal that this approach should be reserved for three categories of women:

> Patients who categorically refuse surgical intervention, those who have undergone local excision of a tumour that was mistakenly considered benign and then referred by their surgeon for post-surgery radiotherapy, and those to whom this form of treatment was deliberately proposed by their surgeon.[5]

The third category included, one may assume, women whose doctors were in favour of conservative therapy for breast cancer, which in the 1960s constituted a small but growing segment of cancer experts.

Baclesse started his experiments with the treatment of breast tumours by radiation alone in 1936. The Curie Institute (known at that time as the Curie Foundation of the Radium Institute) would have been an ideal place for these experiments. The Radium Institute doctors were among the pioneers of radiotherapy of cancer worldwide. They had developed more efficient ways of delivering radiation and applied them to the treatment of head and neck malignancies and uterine tumours. The local medical culture might have favoured the development of radiotherapy of breast tumours. This did not happen, however. Surgeons at the Curie Foundation advocated 'heroic surgery' for breast cancer and systematically promoted extensive operations for women with the disease. This essay attempts to unravel the history of a co-existence between these seemingly contradictory therapeutic trends, and examine the meaning of each for patients.

Physicians who worked at the Curie Foundation, an institution at the cutting edge of therapeutic innovation, often believed that patients should accept 'everything' (that is, harsh mutilating therapies) in exchange for the promise of cure. On the other hand, these physicians were aware that not all their patients accepted this view. Baclesse's published experiments with non-surgical cures for breast cancer were conducted on two very different populations. The legitimate targets of such attempts were women diagnosed with breast cancer who from the start had been classified as 'inoperable' by their surgeons and those who suffered from a recurrent, inoperable tumour.[6] Notes on cancer patients treated by radiotherapy alone between 1919 and 1939 indicate that in the great majority of cases, this treatment was proposed to patients who could not be helped by surgery and were thus seen as incurable. Radiotherapy was considered a purely palliative approach, although, as Baclesse and other radiotherapists had shown, it occasionally induced long-term remissions.[7] In parallel, some women with borderline tumours (growths that could not be clearly classified as malignant or benign) were treated with X-rays alone. The rationale given for such treatment was that the tumour was probably benign, but it might become malignant later. Radiotherapy in such cases was therefore seen as a preventive rather than curative approach.[8] Moreover, it is possible that women who learned that the experts were not sure if their tumours were truly malignant, were more reluctant to agree to radical surgery, and their doctors were more reluctant to push them in this direction.

Starting in the 1940s, Baclesse published a series of articles on the treatment of breast cancer with radiotherapy alone. These articles dealt nearly exclusively with the treatment of inoperable tumours. Meanwhile, however, he also tried this method on 'illegitimate' cases, that is, women whose breast cancer was classified as operable but who refused to be mutilated and asked for alternatives. His 1960s paper was the first publication focused exclusively on such 'illegitimate' cases. We have seen that Baclesse was reluctant to recommend the generalisation of this method. Nevertheless, he explained that:

> we have the duty to answer the anguished interrogations of patients who categorically refuse all mutilating surgical intervention and want radiotherapy. We believe today that when faced with a formal refusal of a patient to submit herself to an extensive surgical intervention, we should consider seriously the possibility of a tumourectomy followed by irradiation, especially if the patient is over 35.[9]

Studies looking at the patient's role in the promotion of conservative treatment of breast malignancies usually suggest that women's resistance to radical mastectomy started in the 1970s.[10] However, I will argue that, at least in France, the rigid adherence to the principle of radical surgery in each case of 'operable' breast cancer was eroded much earlier. The combined effect of the reluctance of some doctors to apply this principle to 'borderline' or uncertain cases (a category which included a significant proportion of 'early', that is, small and localised breast tumours) and the resistance of 'unruly' patients to consent to mutilating surgery, favoured the search for alternative therapeutic solutions.

## Radiotherapy: An alternative to surgery?

The Radium Institute, the precursor of the Curie Institute, was created thanks to an important legacy given to the Pasteur Institute.[11] Inaugurated in 1912, the Institute represented an alliance of two institutions: Marie Curie's physics laboratory, which was affiliated with the Science Faculty of the University of Paris, and a laboratory for the study of biological effects of radiation, the Pavillon Pasteur, affiliated with the Pasteur Institute. The director of the latter, Emile Roux, named Claudius Regaud (1870–1940), a Lyon physician who studied the physiological effects of radiation, as the director of the biology laboratory of the Institute. Regaud was soon joined by his former collaborator from Lyon, Antoine Lacassagne.[12] At first, the new laboratory was mainly dedicated to fundamental research. Regaud's experience as the head of a medical radiology unit during the First World War led him, however, to focus also on the medical applications of radium and X-rays. After the war, Regaud brought some of his collaborators at the Army's medical radiology unit to the Radium Institute, among them Henri Coutard (1876–1949), who became the leader of the Institute's school of X-ray therapy.[13]

Coutard, like Regaud, was at first interested in biological effects of radiation, a topic he studied (from 1912 to 1914) in Jean Danne's laboratory in Gif sur Yvette, near Paris. When he was recruited to Regaud's radiology unit during the war, he began investigating the medical applications of radiation. In 1920, Marie Curie and Claudius Regaud received an important donation from one of the pioneers of radiotherapy in France, Dr Henri de Rothschild, for the promotion of therapeutic activities at the Radium Institute. This donation, named the Curie Foundation, enabled the Institute to purchase radium and X-ray equipment and to open a dispensary dedicated to the outpatient treatment of cancer.[14] The combination of fundamental research (conducted at the Pavillon Pasteur) and clinical applications (conducted at the dispensary) favoured the development of innovative approaches to the therapy of malignant tumours.[15] Regaud and Lacassagne improved the use of filtration so that X-ray treatment could be used to treat more profound malignant lesions. At the same time, a more accurate quantification of the amount of radiation delivered to patients increased the efficacy of therapies and reduced their side effects.[16]

Coutard is associated with two important innovations made at the Radium Foundation: the fractionation of radiation (the administration of lower doses of radiation over longer periods of time), and the evaluation of the biological effects of radiotherapy through careful observation of the effects of radiation on skin and mucous membranes. These two elements were the basis of the 'Coutard method', which was celebrated as an approach that could cure cancers previously perceived as invariably fatal, especially head and neck tumours. The cure rates obtained by Coutard were not very impressive in absolute terms (from 15 per cent to 30 per cent, depending on the tumour), but the very fact that some patients with these tumours did survive was hailed as a miracle. By the 1920s, Coutard had reached a quasi-legendary status among radiotherapists, partly thanks to his clinical successes, and partly because of his colourful personality. His method for treating head and neck tumours was, however, difficult to transfer to other places, as his approach's most important elements were non-transmissible clinical skills and an intensive, highly personalised care. Radiotherapists were expected to examine every single patient immediately before and immediately after a radiotherapy session and to carefully observe localised and generalised reactions to radiation. Changes in skin and mucous membranes, Coutard proposed, paralleled changes in the tumour, and radiotherapists thus needed to learn 'how to perceive and interpret subtle modifications of head and neck tissues'.[17] At the same time, Coutard together with Regaud developed a protocol for the X-ray treatment of cervical cancer. This approach induced more severe complications than radium therapy, and was usually reserved for more advanced cases where the extent of the malignant lesion made a local application of radium difficult. Among 31 patients treated by this method between 1920 and 1925 at the Curie Foundation, six developed severe side effects that led to the interruption of the therapy. On the other hand, the treatment did help some otherwise incurable patients: seven out of the 25 women with advanced cervical cancer who completed the entire treatment cycle entered long-term remission.[18]

Coutard worked at the Radium Foundation until 1937, when he left France for the US.[19] He was replaced by his student and protégé, François Baclesse (1896–1967). Baclesse, who arrived at the Curie in 1926, was seduced by the therapeutic approaches developed by Coutard, and further extended their uses.[20] He also specialised exclusively in X-ray treatment, leaving radium therapy to his colleague, Simone Laborde. Laborde, a physician married to Pierre Curie's student and close collaborator, Albert Laborde, had become interested in the medical uses of radium before the First World War. Named head of the 'curietherapy' department (the local name for therapeutic uses of radium), she specialised in the treatment of cervical and uterine cancers. Laborde was less inclined than some of her colleagues to use 'heroic' therapies that caused irreversible damage to healthy tissues. The goal of curietherapy, she stressed, was to eliminate a malignant tumour, not to induce a large-scale destruction of normal cells.[21] On the other hand, Laborde, like her colleagues at the Curie Foundation, viewed many of the secondary effects of radiotherapy as normal. Notes made during follow-up visits in patient records in the 1920s and 1930s display a contrast between the doctor's overall evaluation: 'the patient is well' (*'la malade va bien'*) and an appended list of her symptoms: pain, scarified skin, digestive and urinary problems, insomnia, difficulty in maintaining certain bodily positions, and reduced mobility.[22]

In the early 1920s, radium therapy of uterine cancers at the Radium Institute was systematically combined with hysterectomy. Patients were first treated with radium, either by external curietherapy or by internal insertion of radium tubes, and then underwent surgery. By the mid-1920s, though, the Curie Foundation's physicians had arrived at the conclusion that radiotherapy alone – usually a combination of external radiation (X-ray or radium) with intra-cavital radium therapy, but in some cases external radiation therapy only – was as effective as the combination of radiotherapy with surgery.[23] Improved radiotherapy techniques in the late 1920s and 1930s led to a marked improvement of survival rates of patients with stage I cervical cancer: as many as 60 to 70 per cent of these patients were alive five years after the diagnosis.[24] Unfortunately, the majority of patients seen at the Foundation had more advanced tumours (stage II and beyond), and thus much lower survival rates – from 20 to 25 per cent at five years. While radiotherapy failed to cure the majority of these patients, it often alleviated pain and other distressing symptoms. On the other hand, in some cases the secondary effects of radiation therapy – extensive scarification, infections, severe urinary and digestive complications, ulcerations, and pathological fractures – turned out to be as bad as or even worse than the disease itself. The Curie Foundation radiotherapists insisted therefore on the key importance of skilled, labour-intensive supervision of all the radiation treatments.[25]

Baclesse improved the methods of treatment of cervical cancers with X-rays alone, and had shown that a careful execution of these protocols reduced the risk of severe complications. His success in this domain might have stimulated his interest in the treatment of breast cancer with X-rays.[26] His early experiments with the irradiation of breast tumours were presented as attempts to reduce the size of the tumour before radical surgery (that is, as an adjuvant therapy). His first study on this subject, made in collaboration with the surgeon André Tailhefer and the cytologist Georges Gricouroff and published in 1939, showed that in some – but not all – cases a breast tumour could be destroyed by radiation. Twenty-one women underwent X-ray treatment before mastectomy, and after surgery pathologists examined the amputated breast. In two of these 21 patients the tumour's size remained roughly the same; in ten it shrank markedly following the irradiation; in five it disappeared almost totally, and in four pathologists were unable to find any traces of malignant cells remaining in the breast tissue. Baclesse and his colleagues concluded that while radiotherapy could not replace breast surgery, the combination of the two techniques opened interesting therapeutic avenues.[27]

The demonstration that a breast tumour could be totally 'sterilised' by radiation alone encouraged Baclesse to use this technique in selected patients with inoperable breast tumours.[28] In the late 1940s he reported that from 1936 to 1945, 145 women diagnosed with locally extended breast cancer had been treated with this approach. The radiation treatment, Baclesse warned, was long, and needed to be executed with the greatest care, or it could produce severe secondary complications. When this form of treatment was delivered at the Curie Foundation, the same physician was present at every radiation session, carefully observing changes in the patient's tissues and other clinical signs. The need to provide individualised supervision of patients greatly increased the cost of this therapy. However, when done correctly, the therapy worked: the tumour disappeared and no malignant cells were found in the breast during a follow-up biopsy. Alas, the clinical results were less impressive, as 32 per cent of the patients survived three years after therapy and only 17 per cent survived as long as five years. Radiation successfully eliminated the primary tumour, but even so, patients died from metastatic disease. The results of radiotherapy, Baclesse concluded, were thus similar to those given by Halsted's surgery for advanced tumours: both methods successfully prevented local recurrence but were unable to stop the progress of distant metastases. Baclesse also noted – without providing further details – that he obtained much better results in ten women with stage I cancer treated with radiation alone, as nine of these patients were still symptom-free three to five years after treatment.[29]

If radiotherapy was as effective – or as ineffective – as radical mastectomy, but more labour intensive and costly, why should one choose this method? The answer, Baclesse stated in 1952, was very simple: because many women wanted it:

> [T]he patient is a woman, and this is her breast. It is interesting to see the level of courage and persistence a woman is ready to display in order to keep her breast. Moreover, this is true for all women, not only young or middle-aged ones.[30]

An additional advantage was that the treatment usually allowed the patient to lead a normal family life, and, in some cases, to continue to work. The lower social cost of this therapy, Baclesse concluded, probably compensated for its higher monetary cost.[31]

In 1959, Baclesse summarised his results with radiotherapeutic treatment of 310 women with advanced breast tumours. The survival rates – 32 per cent at five years and 15 per cent at ten years – were again similar to the survival rates given for women who underwent radical surgery for advanced breast cancer.[32] The majority of Baclesse's patients received very high doses of radiation, and his drastic treatment was indeed comparable to Halsted's mastectomy. The high radiation doses meant that the cosmetic results of the treatment were often not very good (permanent shrinking of tissues, darkened and hardened skin) and, in spite of careful supervision during therapy, many patients suffered from severe secondary effects: teleangiectasia (visible enlargement of small blood vessels), burns, necrosis of the skin, post-radiation oedema and sclerosis. These latter complications made post-surgery surveillance especially difficult. Some of Baclesse's patients were operated upon later for a suspected recurrence which proved to be false, while others had to be operated for genuine local recurrences. Furthermore, surgery on radiation-damaged tissues was a difficult endeavour: the process of scar formation and healing in heavily irradiated tissues took an exceptionally long time, patients often developed severe post-surgery complications, and thus they usually stayed in the hospital two to three months after surgery.[33]

Baclesse retired in 1961. Between 1960 and 1964, his colleagues at the Curie Institute continued to treat stage I and II breast tumours with radiotherapy alone. They employed lower doses of radiation and therefore limited the severity of secondary effects. The clinical results they obtained with this modified approach, however, were mixed. In 30 per cent of the cases treated in this way, the cancer returned locally and patients needed mastectomies. The choice seemed to be between a course of intensive radiation with a high level of local complications, or a less drastic treatment that carried with it an increased risk of local recurrence.[34] The Curie

Institute physicians decided therefore to abandon this approach and to treat localised breast tumours with a combination of conservative surgery (lumpectomy, sometimes extended) and radiotherapy, reserving radiotherapy alone for exceptional cases only.[35]

## Surgical activism at the Radium Institute

The first surgeon attached to the Radium Institute, Jean Louis Roux Berger (1880–1957), was described by his colleagues and collaborators as an excellent surgeon, skilled and meticulous, but also an open-minded practitioner who, from the very beginning, favoured the association of surgery and radiotherapy.[36] He was also depicted as a cultivated, curious and broad-minded person, a politically progressive man incensed by injustices. Roux Berger was interested in the physiological approach to pathological phenomena, an interest he shared with his close friend, the surgeon René Leriche.[37] However, while Leriche advocated a holistic and conservative approach to surgery, Roux Berger championed surgical activism and led a crusade for the extensive excision of malignant growths, including small, localised ones. An admirer of Halsted, he was persuaded that it was impossible to eliminate a malignant tumour surgically without sacrificing an important amount of healthy tissue. He was also very suspicious of benign tumours and argued that these supposedly harmless growths often had unknown malignant potential, a view which might have stemmed from his physiological approach to cancer.[38]

Roux Berger headed a surgical clinic at the Laennec Hospital, but his main passion was his innovative work at the Radium Institute. Together with Regaud and Lacassagne he pioneered a combination of surgery and radiotherapy at that facility. His main specialty was the treatment of head and neck tumours, but at the same time he became interested in women's cancers. The main problem with the latter, he suggested, was delayed or inappropriate treatment. The culprit was often the woman herself. Roux Berger described a woman who was so afraid to see a doctor that she lived for 14 years with a voluminous tumour in each breast. Another woman hid an ulcerating tumour in the breast from her family for many years. Such cases, Roux Berger stressed, ought to be viewed as consequences of mental pathologies. In other cases, however, he held physicians responsible for inexcusable delays in the treatment of malignancies. Many doctors failed to understand that every tumour of the breast, including those that looked benign, should be viewed as a 'terrain' that favoured cancerous evolution and therefore be promptly excised and examined by a competent pathologist. If a surgeon had the slightest suspicion of malignant or pre-malignant growth, he should proceed immediately and operate. Similarly, any hint of a possibility of cervical cancer was to be taken very seriously. As for breast tumours, the slightest suspicion of a pre-cancerous condition of the cervix should trigger either a hysterectomy ('radical surgery will bring a radical cure') or alternatively, radiotherapy.[39]

Cervical cancer, Roux Berger believed, could sometimes be cured by radiotherapy alone, because only few regional lymph nodes were associated with the cervix. This approach was much less effective in breast cancer, as the number of lymph nodes (in Roux Berger's terms 'lymphatic terrain') directly connected with the breast was extraordinarily large.[40] One of the greatest risks in the therapy of breast cancer, Roux Berger warned, was 'incomplete surgery'. Early in his career he analysed 51 cases of breast cancer surgery, finding that 30 patients suffered from a local recurrence in the first year after their surgery. This sad result reflected the, in his view, erroneous belief of some surgeons that the scope of the operation should be proportional to the size and the perceived gravity of the tumour:

> [S]uch a view of surgery for breast cancer is a real disaster. It encourages the practice of these limited operations, incomplete and insufficient, which, together with too long delay before surgery, are the principal cause of rapid local recurrences.[41]

Unfortunately, Roux Berger added, the general public tended to favour surgeons who were reluctant to conduct extensive operations and had an expectant attitude. Moreover, he argued, some people were discouraged by the observation that even when done correctly, an extensive operation for breast cancer often failed to save the patient's life. Roux Berger, like other surgeons of the time, emphasised the prevention of local recurrences of breast cancer. People who criticised surgeons because they were unable to cure their patients with radical breast surgery, he explained, did not understand that the advent of distant metastases depended on constitutional traits of the patient's body, that is, on elements beyond the surgeon's control:

> [I]f, in cases with similar symptoms, surgery of cancer yields very different results, this is due to differences in the properties of the lymphatic system, its infinite complexity, its extraordinary sensibility.[42]

Another major problem, Roux Berger explained, was the difficulty of making clear-cut distinctions between benign and malignant growths.

Often benign tumours of the breast contained areas of malignant or premalignant transformation. If the woman was older than 45, he recommended radical surgery in all the cases where malignancy was suspected, however slightly. Surgeons might hesitate to operate on borderline or doubtful cases, he pointed out, but studies at the Mayo Clinic indicated that in 62 per cent of such cases, the growths were in fact malignant or pre-malignant:

> [I]t is therefore absolutely necessary to treat all suspicious tumours as cancers. One should not hesitate before a mutilation that, in 38 per cent of the cases will be excessive, but in 62 per cent of the cases will eliminate true cancers and increase the chances of cure. This is perhaps a crude rule, but it seems to be the only one that takes into account the insufficiency of our diagnostic means and the danger of an incomplete surgery for a true cancer.[43]

After his retirement, Roux Berger was succeeded by his assistant, André Tailhefer (1896–1963). Tailhefer had trained with Roux Berger at Hôtel Dieu. Initially he collaborated with Roux Berger on perfecting the surgical treatment of head and neck cancers; later he began to specialise in breast cancer surgery. Tailhefer operated on more than 3000 women, and in the post-Second World War era was considered to be the most experienced French surgeon in this domain. He favoured the more extreme variation of Halsted's mastectomy and developed an extended version of this operation that included the removal of the internal mammary lymph nodes.[44] Tailhefer liked to cite the creed of his teacher, Cuéno, that the largest surgical excisions should be reserved for the least advanced cancers.[45] The only effective way to deal with breast cancer, he stressed, was drastic surgery:

> [W]e are convinced that the operations are nearly always too restricted. The extent of skin and muscular sacrifice is of little importance, the conservation of the great dorsal nerve is of little importance; when the operation is sufficiently large, skin recurrences *in situ* are very rare.[46]

Tailhefer's surgical radicalism extended to the preventive treatment of cancer. Like Roux Berger, he was in favour of the surgical elimination of any suspicious lesions. Speaking about the possibility of degeneration of mastosis (inflammation of the breast) and its potential transformation into malignancy, he explained that such degeneration was probably rare.

> Nevertheless, after this concession to statistics, let us look at this problem from a more practical, and possibly also more humane point of view. We cannot deny the possibility of a cancerous degeneration of mastosis. We should do anything possible to prevent it by appropriate treatment. If the mastosis is localised, such a treatment will be a limited surgical excision. A mild surgical intervention will therefore protect the patient. The low probability of a malignant degeneration is not an important consideration for the patient who may become its victim.[47]

Tailhefer believed that radiotherapy was a useful addition to breast cancer surgery but that its use should be limited to the treatment of advanced tumours.[48] Highly respected at the Curie Foundation, his biographer claims that he was able to persuade his peers about the absolute necessity of radical mastectomy for all 'operable' patients.[49]

## Surgical activism in practice: Spaces of negotiation

Until the 1960s, the Curie Foundation's surgeons seemed to possess full control over the treatment of breast cancer. Reading their theoretical contributions, one might assume that, at least with regard to breast cancer, their approach was comparable to that promoted at the same period by US surgeons.[50] On the other hand, Roux Berger and Tailhefer could not totally disregard the professional environment in which they worked: the professional culture of the Curie Foundation with its focus on radiotherapy, and the fact that since the 1940s French gynaecologists had been favouring medical solutions to gynaecological problems.[51] They agreed with the Foundation's experts that the optimal treatment for uterine cancer was a combination of external and internal radiotherapy, and that hysterectomy should be reserved for exceptional cases. But breast cancer, they argued, was a very different disease, which had to be treated with radical, or even ultra-radical, surgery. Still, Roux Berger and Tailhefer's affirmation that such surgery should be reserved exclusively for operable malignant breast tumours left room for debates among surgeons and oncologists over the precise meaning of the terms 'operable' and 'malignant'.

Many tumours were classed as 'inoperable'. This category included cancers that had already spread beyond the breast at the time of diagnosis, extended or ulcerating cancers, and recurrences after mastectomy. Such tumours were treated with limited surgery and radiotherapy or radiotherapy alone. The goal of 'salvage therapy' was not to produce cures but to reduce suffering. However, it did induce prolonged remissions in some patients. In one case, radiation therapy led to the reclassification of a tumour, from a breast tumour infiltrating the skin (viewed as incurable) to skin cancer (a potentially curable disease): Mme H., a 56-year-old woman, was diagnosed in September 1926 with a breast epithelioma that infiltrated her skin. Her tumour was judged by her doctors too advanced for surgery. Treated with roentegenotherapy and radiotherapy, she developed persistent local radiolesions with ulceration and necrosis, but was still alive and free of cancer symptoms in 1935. A note in her file from September 1936 stated that the original tumour was probably not a breast neoplasm but a hydroadenoma or a hydroepithelioma of the skin. The skin lesions healed slowly, and in April 1941 the patient reported that she was feeling well.[52]

In some cases, the Curie Foundation doctors chose to use radiotherapy to shrink breast tumours prior to surgery. Such radiotherapy occasionally turned out to have diagnostic value. A 33-year-old woman suffered for two years from an intermittent secretion of liquid from her nipple, then noted the presence of a hard lump in her breast. Her physician reassured her that this was a benign cyst. The tumour increased rapidly in size and became painful, and she also began to suffer from pain in her arm. In March 1934, Tailhefer offered a tentative diagnosis of a potentially malignant intracanalicular tumour, and proposed X-ray treatment followed by a radical mastectomy. The patient's doctor was informed that the woman's prognosis was poor, mainly because of her young age. X-ray therapy induced extended radiodermatitis but the size of the tumour remained stable. This was seen as a good sign: malignant breast tumours are usually sensitive to radiation while benign cysts are not. Tailhefer then decided to excise the tumour alone. The histological diagnosis confirmed that the tumour – a fibro-adenoma with intracanalicular proliferation – was not malignant, and the patient's doctor was reassured that the woman's outlook was in fact very good.[53]

Patients treated with radiotherapy before a scheduled mastectomy had time to consult other doctors and hear additional opinions on treatment options. Patients' files often contain letters from family physicians asking for details of diagnosis and prognosis.[54] The time lag between diagnosis and surgery also allowed some women to rethink their willingness to undergo radical surgery. Mme M. L., a 50-year-old woman diagnosed with a breast cancer and enlarged lymph nodes in July 1939 was scheduled to undergo radiotherapy followed by a mastectomy. When her radiation treatment ended in early September 1939, the tumour and lymph nodes were no longer palpable. The patient then announced that she had decided to cancel her surgery because of the international situation.[55]

Until 1939, doctors at the Curie Foundation used frozen section (a biopsy technique that allowed the rapid pathological analysis of tumours during breast cancer surgery) only in exceptional cases.[56] Often the Foundation's doctors declared tumours malignant on the basis of clinical findings alone. Women could then agree to the proposed treatment, negotiate its details, or look for a different physician. When in doubt, Curie Foundation doctors excised the tumours and sent them to their pathology laboratory for a paraffin section, a fixation, cutting and staining technique that provided a more secure diagnosis of malignancy than a frozen section, but took several days. Surgical notebooks of the Curie Foundation for 1923–1936 reveal that in spite of their programmatic declarations in favour of radical surgery for borderline breast tumours, the Foundation surgeons were not always eager to operate on doubtful cases.[57] For example, in March 1925, Mme B. M., a 53-year-old woman with an ulcerating nipple and enlarged axillary ganglion was tentatively diagnosed with Paget's disease of the nipple, a pathology usually classified as malignant. She underwent a tumourectomy, so that her doctors could decide if she should be treated by a radical or a simple mastectomy. The cytological analysis did not find malignant cells. Her doctors observed, moreover, that her biopsy scar healed rapidly and the enlarged ganglion shrank markedly, indicating that the growth was perhaps benign, and the ganglion was enlarged as a result of an inflammation. At that point, they considered abandoning their original plan and instead refraining from additional surgery. Eventually, taking into consideration the patient's age, they decided to perform a simple mastectomy.[58]

Women with suspicious changes in their breasts who were treated at the Curie Foundation were not obliged to sign a blank agreement for a radical surgery, and – unlike patients in the US, where 'one stage' operations were the rule – usually they were not sent to sleep without knowing whether they would wake up with a small scar or an extensive mutilation.[59] They could therefore negotiate with their doctors knowing what the medical recommendations were. Such negotiations were probably shaped by the availability of a variety of approaches to the treatment of breast cancer in France. Women dissatisfied with their doctors' advice could seek out another specialist. Roux Berger and Tailhefer strongly condemned surgeons who did not adopt their radical approach and who agreed to limited surgery for a small or 'borderline' tumour, but they were probably unable to completely ignore the fact that many women preferred the more conservative approaches employed by some of their competitors.

## Radiotherapy for women's cancers in New York City: The 'French method' abroad

The French school of radiotherapy had a major advocate in the US: Maurice Lenz (1890–1974). Lenz had studied medicine at Columbia University, and completed his training in Germany, where he first learned about radiotherapy of cancer. Impressed by the achievements of French radiotherapists he studied in 1924 and 1925 at the Radium Institute. Back in New York, in 1930 he was appointed head of the radiotherapy laboratory at Presbyterian Hospital, which was affiliated with Columbia University. Lenz became an enthusiastic promoter of French therapeutic methods in North America. He corresponded with Regaud, Coutard and Baclesse (the latter was also a personal friend), visited the Curie Foundation regularly, sent his students to study radiotherapy in Paris, and invited French experts to give lectures in the US.

Lenz was familiar with Baclesse's attempts to treat operable breast cancer with radiotherapy alone, but was not persuaded that this was a valid curative approach.[60]

> I have many times emphasised my belief that cancer in the breast is less radioresponsive than the same cancer in lymph nodes. This was taught by Regaud in 1922. In spite of the largest doses which Baclesse and I gave to the breast, there were many cases in which persisting cancer was found microscopically. (...) We do not have a level of dosage at which we can be certain that all cancer cells within the breast will be destroyed. Mastectomy, therefore, is more reliable than radiation for this site.[61]

Lenz treated women with inoperable breast tumours by salvage radiotherapy. This treatment was occasionally effective, but at the price of severe complications. For example Lenz described the case of B. L., a 63-year-old woman admitted with a very large breast tumour and an enlarged axillary node found to be positive for cancer cells. Classed as inoperable, she received high doses of radiation to the breast, the axilla and the supraclavicular region:

> The patient now presents marked skin atrophy, indurations over the most intensively irradiated portions of the breast and a moderate amount of radiation pulmonary fibrosis. While the indurate subcutaneous tissues may contain cancer, there is no sign of activity, and it seems fair to assume that the cancer, temporarily at least, has been controlled by X-ray therapy. Perhaps the cost in tissue change, which in time will undoubtedly get worse, has been too great.[62]

For Lenz, as for many of his colleagues, the choice in breast cancer treatment was not between surgery and radiotherapy, 'knife versus rays', but rather about how to combine these two approaches in order to intensify the destruction of malignant cells. Perfecting existing therapies, Lenz argued, would lead to a more efficient elimination of cancer cells, and therefore to better cure rates. Not all his colleagues were as optimistic as he was. Lenz's former student, Franz Buschke, a radiologist at the UCSF Medical Center, wrote to Lenz in 1957 to ask advice about the organisation of a radiology course, and added:

> in my own mind, I am still thoroughly confused about the indications for Roentgen therapy of carcinoma of the breast, and I will greatly appreciate some fatherly advice (...) I suppose that you have seen Ian MacDonald's paper in the proceedings of the Third National Cancer Conference, 1957. I thought it was very interesting and it thoroughly fits my philosophy. If his observations stand the test of criticism, I think that we should, once and for all, stop our insistence on proving statistically the superiority of one method over the other. If 25 per cent have a favourable prognosis, regardless of what is done, and 50 per cent have an unfavourable prognosis, regardless of the type of therapy, and only in 25 per cent treatment is critical, I do not think that one can ever prove the superiority of one method over the other.[63]

Buschke was impressed by the suggestion by Ian MacDonald, a Californian surgeon, that the trajectory of a malignant tumour was predetermined by its biological traits, and the unique dynamics of host-tumour interactions.[64] Doctors, Buschke hinted, might have strong convictions about the efficacy of the treatments they administered, but, at least in the case of breast cancer, it was not clear if their interventions made as much difference as they liked to believe.

## Groping in the dark: Therapeutic uncertainty and therapeutic choices in oncology

Cervical cancer was seen as a relatively straightforward pathology. Women with small localised, cervical lesions could often be cured; those with larger and more extensive tumours usually died a few years at most after their diagnosis. The main problem with this malignancy stemmed from the fact that when a woman noticed gynaecological symptoms, her cancer often had already spread, at least locally. Breast cancer was much less predictable. Some patients with extended cancerous lesions lived for a long period of time, while others, diagnosed with small, localised tumours, died rapidly from distant metastases. Experts energetically preached the gospel of 'do not delay', but those who worked with breast cancer patients on a daily basis were aware that the rule – that a rapid intervention led to a cure – was far from absolute. Some tumours were unusually aggressive, others were unusually slow growing, and clinical or histological observation did not provide reliable ways of identifying such aggressive or sluggish tumours. Lenz recalled the case of a woman who consulted him about a tumour in her breast and was strongly advised to undergo a radical mastectomy. She rejected this advice and eschewed all treatment until her tumour had completely infiltrated her breast and extended itself to various portions of her skeleton. Only then, 22 years after her initial diagnosis, did the pain of bone metastases force her to agree to undergo radiotherapy.[65] Lenz's collection of case studies contained a wealth of such atypical cases:

L. S. C. 29 years. Trained nurse, seen in November 1929. Lump 5 to 6 months. Initial diagnosis was cystic disease. Operated two days later, cancer with metastases. 5 nodes involved. Received postoperation radiation from 1930–1932. Serious strep infection of the arm following surgery. In spite of extensive lymph node metastases and an inexcusable delay after discovery, the patient is still alive after 6 years.

A. E. 44 years. Seen in November 1917, after 1 1/2 year delay. X rays seem to show bone metastases, but she was operated nevertheless and given postoperative radiotherapy. The patient suffered from severe side effects of radiotherapy and bad electric pad burns. Alive 17 years later, in spite of a very poor diagnosis.

H. T. 61 years. Was seen in November 1930 and operated on 3 days later. Frozen section revealed 2 positive lymph nodes out of 15. In November 1931 her doctors stated extensive metastatic lung involvement. She was readmitted in October 1932, cachetic and febrile, for what was thought then would be terminal care, but she unexpectedly improved. In May 1934 pulmonary symptoms receded. A radiation pneumonitis might have been mistaken for a metastatic cancer. The patient is alive and well 6 and half years after her surgery.

E. W. R. a black woman [no age stated]. Seen in April 1918, with large lump in the breast and numerous positive lymph nodes. Underwent radical mastectomy. In November 1935, she was diagnosed with a suspicious mass in contra lateral breast. The patient did not agree to an operation; she trusts in a sect called 'Silent Unity' and in the efficacy of prayer. In May 1936, the lump in her breast became less noticeable with 'Silent Unity treatment!' The patient is alive and seems to be well.

R. S. 44 years. Seen in November 1923. Year earlier she discovered a lump in her breast about 2cm, with small areas of induration. Clinical diagnosis was uncertain. She underwent radical mastectomy, and the histological verdict was 'carcinoma with many metastases'; cystic disease with 'dangerous' type of epithelial hyperplasia. Axillary glands 'full of carcinoma'. The prognosis was very poor, but the patient is alive 13 years later.

M. S. 49 years. Seen in December 1917, with a lump size of hen egg in her breast. Mastectomy revealed carcinoma with metastases. March 1918, she underwent subcutaneous mastectomy of left breast, because of suspicious areas of indurations but histological examination did not reveal carcinoma. The patient, who would be considered for many inoperable and who showed extensive axillary involvement, is alive and well for 19 years.[66]

Lenz concluded that over-representation of either slow-growing or rapidly-growing tumours in a small series could introduce strong biases and lead to inaccurate conclusions about the effectiveness of a specific therapy. He believed that a better collection of pooled data would lead to the identification of truly efficient treatments for breast cancer.[67] Radiologists at the Curie Foundation had a more reserved view of treatments for this disease. In 1932 they summed up available data on outcomes of therapies for breast cancer and concluded that the combination of radiotherapy and surgery seemed to confer a modest survival advantage. They added, however, that the majority of authors only provided data about survival at three, or at best five years after diagnosis. In breast cancer, they added, the only meaningful data were those that reported survival for ten years or more, and these data – where they existed – indicated that the 'true' cure rate for breast cancer unfortunately was very low, independently of the method used.[68]

In 1962, the International Task Force on Breast Cancer, based in Bethesda, Maryland attempted to assess the efficacy of different approaches to the therapy of breast cancer on the basis of data from cancer registries in six countries: Denmark, Finland, France, Norway, the UK, and the US. The coordinators of this investigation decided to divide breast cancer cases into two groups: tumours limited to the breast with no axillary involvement (Group I) and all other breast malignancies (Group II); they then compared survival at five years.[69] The study demonstrated important differences in the choice of therapies among the surveyed countries. In the US, 90 per cent of women diagnosed with Group I breast tumours were treated with surgery only, while in Norway 90 per cent of women diagnosed with such tumours were treated with a combination of surgery and radiotherapy. In the UK 28 per cent of Group I patients were treated with surgery alone and 60 per cent with combined treatment; in Finland, 18 per cent with surgery alone, and 75 per cent with combined therapy; in France, 30 per cent with surgery alone, 44 per cent with combined treatment, and 22 per cent were treated with radiotherapy but not surgery. France was the only country where some women with localised breast cancer were treated with radiation alone. The therapy of group II tumours displayed similar variability. Sixty-four per cent of the Group II patients in the US were treated with surgery alone, as compared to 20 per cent of the British and 10 per cent of the French patients.[70] But in spite of the important differences in therapeutic choices, the results of therapy were surprisingly uniform. In all six countries, between 80 and 85 per cent of the women with Group I cancers and 40 to 45 per cent of women with Group II cancers were alive and symptom-free five years after their initial diagnosis.[71]

In 1979, Maurice Fox analysed breast cancer statistics with regard to the outcomes of radical versus conservative therapies. He found that while the incidence of breast cancer in the US had increased greatly, especially from the 1960s on (probably as a result of the introduction of mammographic screening), the mortality from the disease remained stable. Fox concluded – like other cancer experts before him – that the trajectories of breast cancer patients were shaped by unknown biological variables:

> [A]nalysis of survival curves of women with breast cancer suggest that two or more populations exist, with about 40 per cent suffering fatal outcome, unaffected by treatment. The remaining 60 per cent exhibit a relative mortality only modestly different from that of women of similar ages without an evidence of disease. Increasing detection of an entity that is histologically defined as malignant but biologically relatively benign could account for the observed increase in incidence.[72]

Fox added that 'as a biologist who had analysed existing clinical and biological data, I'm puzzled as to why so many physicians continue to select more radical forms of intervention'. It would be a great disservice to physicians and patients, he insisted, to interpret his observations as an invitation to refrain from therapy. In the absence of more effective diagnostic means, each clinically manifest breast cancer (meaning: a lesion diagnosed as a cancer by a pathologist) should be treated as though it was biologically cancer (a deadly disease). At the same time, doctors and patients should be aware of the existence of a very problematic situation: 'recognition that a dilemma exists is a prerequisite to progress in this important and troubling field'.[73]

Fox was a microbiologist who mainly specialised in the study of mutations in bacteria. It may be easier to reflect on the moral and practical dilemmas of cancer treatment far away from a responsibility for treatment of the sick. A practitioner confronted daily with the suffering of his patients may be more inclined to transform clinical uncertainty into a blueprint for action. André Tailhefer, all his surgical radicalism notwithstanding, was perfectly aware of the high level of uncertainty in the treatment of breast cancer. He knew that doctors were unable to cure aggressive forms of this malignancy and that experts were unable to evaluate whether a given tumour had already spread beyond its initial localisation.[74] The clinical evidence, Tailhefer explained, was not sufficient to make therapeutic decisions. However, he continued, this did not mean that doctors should favour an expectant attitude or conservative solutions. Just the opposite was true: a high level of uncertainty legitimated surgical radicalism:

> The use of prophylactic or systematic therapies is not 'acknowledgment of impotence' but, on the contrary, is grounded in a very precise knowledge, the one of the extent of impossibility of evaluation.[75]

Using a more conservative approach for the therapy of small, localised tumours, Tailhefer argued, might rob numerous cancer patients of the chance to achieve a cure.[76] The key word here was 'chance'. Such a chance operated on two levels: the chance that a small, apparently localised tumour was in fact more aggressive than it appeared, and the chance that aggressive surgery would improve long-term survival.[77]

Lenz held similar views. In his notes, he discussed at some length a case of a young woman diagnosed with an intraductal papillary growth (a benign tumour) in 1923. The growth was removed, but in 1930 she developed a different lump in her breast. Suspected of harbouring a true malignancy she underwent an exploratory dissection of axillary lymph nodes. The dissection did not find any signs of malignancy. Nevertheless Lenz recommended radical surgery:

> The only question that arises is whether there should be a radical operation or simple mastectomy. In this instance I think it is wise to do a radical mastectomy, because of the lack of faith I have in my ability to be sure, even after the breast has been removed, how to classify such cases in regard of malignancy and that it is by all odds best to give the patient the benefit of the doubt.[78]

Quantifying uncertainty, Ian Hacking has shown, 'tamed chance', and made possible the transformation of ignorance into a scientific approach.[79] In the treatment of female cancers, uncertainty was employed to legitimate therapeutic activism. Such an activism allowed the doctors to feel that they had done 'everything possible', whatever the price for the patient. Some women rejected this logic.

## Notes

1  Baclesse *et al.* 1960. The surgical removal of the tumour alone (a biopsy) was usually undertaken in order to provide a diagnosis.

2   Stage I tumours are small and limited to the breast, stage II tumours spread to axillary lymph nodes, but not beyond them.

3   Baclesse was careful not to speak about cures of breast cancer. He explained that because this disease sometimes returned after a very long symptom-free period, one could at most speak about 'apparent clinical cures': Baclesse 1959a, p. 444.

4   Baclesse *et al.* 1960, p. 139.

5   Baclesse *et al.* 1960, p. 137.

6   On the history of 'salvation' treatment of inoperable breast tumours see Aronowitz 2007, pp. 133–9.

7   'Breast cancer, treated with radiation alone, 1919–1939', Baclesse's notes, Department of Patient Files, Curie Institute Hospital Archives [hereafter: Curie Hospital Patient Files]. I am indebted to the head of this department, Laurence Leclerc, for her interest in historical studies and invaluable help.

8   'Breast cancer, treated with radiation alone, 1919–1939', Baclesse's notes, Curie Hospital Patient Files. For the history of frozen section, see Wright 1985, and for the history of its use in the diagnosis of breast tumours, Löwy 2007.

9   Baclesse *et al.* 1960.

10   Kushner 1975; Altman 1996; Cassamayou 2001.

11   Daniel Iffla-Osiris, who legated his large fortune to the Pasteur Institute, expressed in his testament the wish that part of the money should be dedicated to the study of cancer. Coptes rendus de conseil d'adminstration de l'Institut Pasteur, 20 March 1907, Pasteur Institute Archives.

12   Regaud 1936; Vincent 1997.

13   The early history of radium therapy in France has been studied by Vincent 1999. On the development of radiotherapy in Germany see van Helvoort 2001, and on the history of radiotherapy of gynaecological cancer in the UK, Moscucci 2007.

14   The Curie Foundation's hospital opened in 1936. After the Second World War, the unified institution (the dispensary and the hospital), was renamed Curie Institute.

15   The Curie Foundation had special agreements with several Assistance Publique hospitals and private clinics in Paris, to allow radiotherapy of patients that needed to be hospitalised. The poorest patients were often treated at the Pasteur Institute Hospital. Process Verbaux, Conseil d'Adminstration, Curie Foundation Archives, Curie Institute Museum.

16   Vincent 1999.

17   Baclesse 1950a.

18   Regaud and Coutard 1925.

19   Coutard had a rather turbulent career in the US. He worked first at Chicago Tumor Clinics with Cutler, then at Penrose Tumor Clinics, Colorado Springs, where he dedicated himself to fundamental research and developed rather idiosyncratic theories on the origins of cancer. His later views were interpreted by some as signs of a mental disease, although people who knew him personally denied this. E. R. N. Griegg to Maurice Lenz, 10 March 1973 and Adolf Zuppinger to E. R. N. Griegg, 2 February 1973, Box 2, Folder 3, Maurice Lenz Papers, Columbia University Medical School Archives [hereafter: Lenz Papers].

20   Lacassagne 1968.

21 Huguenin 1946, pp. 166–7.
22 Treatment books: Cervical cancer, Treatment notes, Baclesse, Department of Patients Files, Curie Institute Hospital Archives.
23 Regaud 1926. One of the major drawbacks of hysterectomy was a high rate of surgery-related deaths.
24 Five years survival is a meaningful measure in cervical cancer because late recurrences are rare.
25 Baclesse 1950b.
26 Baclesse 1936; Baclesse and Fernandez-Colmeiro 1942.
27 Baclesse *et al.* 1939.
28 Baclesse also might have been encouraged by reports of cures of breast cancer with radiation alone, e.g. Keynes 1937, Pfahler and Vastine 1938. The term 'sterilised' was used by the Curie Foundation doctors to describe the complete elimination of cancer cells.
29 Baclesse 1949.
30 Baclesse 1952, p. 292.
31 Baclesse 1952, pp. 292–3.
32 Baclesse 1959a and 1959b.
33 Calle and Pilleron 1967.
34 Calle *et al.* 1973.
35 Today, severe complications caused by radiotherapy for breast cancer, such as permanent skin burns and skin necrosis, are found only in a small percentage of women (between 1 and 3 per cent). They are usually attributed to faulty radiation technique. However, the majority of treated women experience mild to moderate skin irritation and teleangiectasia. Cf Turesson *et al.* 1996; Ramsay and Birrell 1995; Clinical Practice Guidelines for the Care and Treatment of Breast Cancer 1997; Huang *et al.* 2002.
36 Wolfromm 1957.
37 On René Leriche and his role in interwar France, see Weisz 1998.
38 de Gaudart d'Allaines 1957.
39 Roux Berger 1921.
40 Roux Berger 1922a.
41 Roux Berger 1922b.
42 Roux Berger 1940.
43 Roux Berger 1924–1925, p. 6.
44 Moyse 1963. Tailhefer was expected to retire at 65 (in 1960), but the director of the Curie Institute asked him to continue to work there, and he was employed by the Institute until his death in 1963. Baclesse retired from the Curie Foundation at the prescribed age of 65, and took a job at the American Hospital in Paris.
45 Tailhefer 1954a.
46 'Nous sommes persuadés qu'on opère toujours trop étroitement. Peu importe le sacrifice cutanée et musculaire, peu importe la conservation du grand dorsal. Les recidives cutanées *in situ* sont l'exception si on opère largement'. Tailhefer 1934, p. 465.
47 Tailhefer 1960.
48 Tailhefer and Dulac 1943.
49 Moyse 1963.
50 On the history of breast cancer treatment in the US, see Lerner 2001.

51 Weisz 2005.
52 Case No. 107, Folder 'Breast Cancer treated by radiotherapy alone', Curie Hospital Patient Files.
53 Patients files, 1930–39, *ibid.*
54 Only 10 to 20 per cent of the Foundation patients were private patients paying for all their treatment. The so called 'charity patients' often appeared educated and articulate, and many had their own family doctor who played an active part in their treatment. They were classed as 'charity patients' mainly because of the high cost of radiotherapy. Process Verbaux, Conseil d'Adminstration, Curie Foundation Archives, Curie Institute Museum.
55 'L'opération chirurgicale envisagé n'est pas accepté par la patiente en raison des événements internationaux'. Case No. 357, File, 'Breast Cancer treated by radiotherapy alone', Curie Hospital Patient Files.
56 By contrast, in the US, frozen section became the state of art technique in surgery for breast cancer in the 1920s: Wright 1985, Löwy 2007.
57 Operation room notebooks, 1919–1939; 'Breast cancer treated by radiation alone, 1919–1939', 'Cumulative data on uterine cancer patients, 1919–1939', Curie Hospital Patient Files.
58 'Reports of surgeries', Notebook for 1924–1925, Case No. 158, Curie Hospital Patient Files.
59 In a few cases, surgeons who believed that they would perform only a tumourectomy, decided during surgery that the lump had the typical gross appearance of a malignant growth and performed a radical mastectomy. 'Reports of surgeries', Curie Hospital Patient Files.
60 Coutard to Lenz, 18 March 1933, Folder 3, Box 2, Lenz Papers.
61 Lenz to Cushman Haagensen, 17 June 1968, Folder 11, Box 11, Lenz Papers.
62 Folder 5, Box 15, Lenz Papers.
63 Bushke to Lenz, 19 August 1957, Folder 12, Box 1, Lenz Papers.
64 MacDonald 1951; MacDonald 1958. Lerner 2001 discusses the debates about the biological predetermination of malignant tumours in the 1950s and 60s. See also Timmermann's chapter in this volume for similar considerations for lung cancer.
65 Undated manuscript on breast cancer [1950s?], Folder 2, Box 11, Lenz Papers.
66 'Breast cancer', Folder 5, Box 15, Lenz Papers.
67 Lenz was a member of the Standardization Committee of the American Radium Society. In 1937 he organised a survey among the Society members in order to gather data on their uses of radiotherapy. Folder 1, Box 12, Lenz Papers.
68 Pierquin and Richard 1939.
69 This ad hoc division adapted by the group was different from the classification adopted by the WHO in 1952, and then refined in 1956. Denoix and his colleagues chose to simplify this classification and make only one distinction between Group I, that is, stage I cancer (localised tumour without lymph node involvement) and Group II, which included stage II, III and IV cancers. On the history of staging cancers see Ménoret 2002.
70 France appears to be the only major country where radiation only is frequently used in the treatment of breast cancer. Thirty-three per cent of French Group II patients, compared to only 6 per cent in the US patients, 9 per cent in the UK, and 5 per cent of Norwegian patients were treated with radiotherapy only.
71 Leguirenas and Gelle 1962.

72 Fox 1979, p. 494.
73 *Ibid.* Fox's suggestion that each clinically manifest breast cancer should be treated as though it was biologically cancer, was by itself problematic. What a 'clinically manifest breast cancer' is remains controversial.
74 Tailhefer 1954b.
75 Tailhefer 1954b, p. 567.
76 'il sera indamissible que nous ne donons pas aux toutes le maximum des chances'. Tailhefer 1954b, p. 567.
77 Lerner 2001 and Aronowitz 2007 discuss why faith in the efficacy of radical mastectomy persisted for so long despite an absence of corroborating data.
78 Folder 5, Box 15, Lenz Papers.
79 Hacking 1990.

## Works Cited

Altman R. 1996, *Waking Up, Fighting Back: The Politics of Breast Cancer*, Boston: Little, Brown and Company.

Aronowitz R. A. 2007, *Unnatural History: Breast Cancer and American Society*, New York: Cambridge University Press.

Baclesse F. 1936, 'Quelques considerations sur la Roengentherapie employée seule dans le traitement des epithéliomas avancées du col d'utérus et du vagin', *Radiophysiologie et Radiothérapie*, 3, 377–418.

Baclesse F., Gricouroff G. and Tailhefer A. 1939, 'Essai de Roentgenotherapie du cancer du sein suivie d'operation large', *Bulletin du Cancer*, 28(5), 729–43.

Baclesse F. and Fernandez-Colmeiro J. M. 1942, 'Quelques remarques sur le traitement radiotherapeutique des adenocarcinomes du col uterin', *Bulletin du Cancer*, 30(2), 118–28.

Baclesse F. 1949, 'Roentgen Therapy as the Sole Method of Treatment of Cancer of the Breast', *American Journal of Roentgenology and Radium Therapy*, 62(3), 311–19.

Baclesse F. 1950a, 'Coutard (1876–1949)', *Journal de Radiologie et d'Electrologie*, 31(7–8), 475.

Baclesse F. 1950b, 'Roentgen Therapy Alone in the Treatment of Advanced Cervico-Uterine Cancer', *The American Journal of Roentgenotherapy and Radium Therapy*, 63(2), 252–5.

Baclesse F. 1952, 'La roentgenotherapie seule dans le traitement du cancer du sein', *Acta Union Internationale Contre le Cancer*, 8(1), 284–93.

Baclesse F. 1959a, 'Cancer du sein: Roentgenotherapie seule', *Journal de Radiologie et d'Electrologie*, 40(8–9), 444–6.

Baclesse F. 1959b, 'Le prognostic eloigné des cancers du sein (stades II, III, IV, à l'exclusion du stade I) traités par roentgenothérapie suele', *Bulletin du Cancer*, 46(3), 594–8.

Baclesse F., Ennuyer A. and Cheguillaume J. 1960, 'Est on autorisé à pratiquer une tumorectomie suivie de radiothérapie en cas d'un tumeur mammarie', *Journal de Radiologie et d'Electrologie*, 41(3–4), 137–9.

Calle R. and Pilleron J. P. 1967, 'Chirurgie du sein après cobaltothèrapie a doses cancericides', talk at the Symposium europeen de radiologie mammaire, Strasbourg, 1–3 July 1966, reproduced in *Journal de Radiologie et d'Electrologie*, 48(11).

Calle R., Pilleron J. P. and Schlienger P. 1973, 'Therapeutique a visé conservatrice des epitheliomes mammaires', *Bulletin du Cancer*, 60(2), 217–34.

Cassamayou M. H. 2001, *The Politics of Breast Cancer*, Washington DC: Georgetown University Press.

Clinical Practice Guidelines for the Care and Treatment of Breast Cancer 1997, 'Breast Radiotherapy After a Breast Conserving Surgery', *Cancer Prevention and Control*, 1(3), 228–40.

Fox M. S. 1979, 'On the Diagnosis and Treatment of Breast Cancer', *JAMA*, 241(5), 489–94.

de Gaudart d'Allaines F. 1957, 'Jean Louis Roux–Berger', *Bulletin de l'Academie Nationale de Médecine*, 28–9, 611–18.

Hacking I. 1990, *The Taming of Chance*, Cambridge: Cambridge University Press.

van Helvoort T. 2001, 'Scalpel or Rays: Radiotherapy and the Struggle for the Cancer Patient in Pre-World War II Germany', *Medical History*, 45, 33–60.

Huang E. Y., Chen H. C., Wang C. J., Sun L. M. and Hsu H. C. 2002, 'Predictive Factors for Skin Telangiectasia Following Post-Mastectomy Electron Beam Irradiation', *British Journal of Radiology*, 75, 444–7.

Huguenin R. 1946, 'L'apport de la France dans l'étude du cancer', in M. Alajaune (ed.) *Ce que la France a apporte à la médecine dès le débout de XX siècle*, Paris: Flammarion, 139–73.

Keynes G. 1937, 'Conservative Treatment of Cancer of the Breast', *BMJ*, 4004, 643–7.

Kushner R. 1975, *Breast Cancer: A Personal Story and an Investigative Report*, New York: Harcourt Brace Jovanovich.

Lacassagne A. 1968, 'François Baclesse, 1896–1967', *Journal de Radiologie*, 48(3–4), 13–15.

Leguirenas J. and Gelle X. 1962, 'Prognostic et évolution du cancer du sein: Etude comparative dans 5 pays', *Bulletin de l'Institut National d'Hygiène*, July–August, No. 4, 569–83.

Lerner B. 2001, *The Breast Cancer Wars: Fear, Hope and the Pursuit of Cure in Twentieth Century America*, Oxford and New York: Oxford University Press.

Löwy I. 2007, 'Breast Cancer and the "Materiality Of Risk": The Rise of Morphological Prediction', *Bulletin of the History of Medicine*, 81, 241–66.

MacDonald I. 1951, 'Biological Predeterminism in Human Cancer', *Surgery, Gynecology and Obstetrics*, 92, 443–52.

MacDonald I. 1958, 'The Individual Basis of Biologic Variability in Cancer', *Surgery, Gynecology and Obstetrics*, 106, 227–9.

Ménoret M. 2002, 'The Genesis of Notion of Stage in Oncology', *Social History of Medicine*, 15(2), 291–302.

Moscucci O. 2007, 'The "Ineffable Freemasonry of Sex": Feminist Surgeons and the Establishment of Radiotherapy in Early Twentieth-Century Britain', *Bulletin of the History of Medicine*, 81(1), 139–63.

Moyse P. 1963, 'André Tailhefer, 1896–1963', *Quotidien de Médecin*, 12 September.

Pfahler G. E. and Vastine J. H. 1938, 'Carcinoma of the Breast', *JAMA*, 110(8), 543–9.

Pierquin J. and Richard G. 1939, 'L'association des la chirurgie et des radiations dans le traitement du cancer de sein', *Paris Médical*, March 19, 250–8.

Ramsay J. and Birrell G. 1995, 'Normal Tissue Radiosensitivity in Breast Cancer Patients', *International Journal of Radiation Oncology, Biology and Physics*, 31, 339–44.

Regaud C. and Coutard H. 1925, 'Les resultats et technique de roentgenotherapie dans les cancers de col d'uterus', *Gynecologie et Obstétrique*, 12(4), October 12.

Regaud C. 1926, *Traitement des cancers de col d'uterus par les radiations*, Bruxelles: Imprimerie Médicale et Scientifique.

Regaud C. 1936, *Notice sur les travaux scientifiques de Claudius Regaud, 1893–1935*, Paris: Presses Universitaires de France.

Roux Berger J. L. 1921, 'Reflexion sur le cancer: Le diagnostic precoce', *Bulletin Medical*, No. 27. Paris: Imprimerie Polyglotte E. Henri.

Roux Berger J. L. 1922a, 'Les therapies associés: Chirurgie, rayons X, radium', *Paris Médical*, 17 March, 923–6.

Roux Berger J. L. 1922b, 'Cinquante et une observations des récidives post-opératoires de cancer du sein', *Bulletin et Memoires de la Societé de Chirurgie de Paris*, meeting of 31 May.

Roux Berger J. L. 1940, 'Cancer et le système lymphatique: Faut il respecter les tissus sains?' *La Presse Médicale*, July 31–August 3.

Roux Berger J. L. 1924–1925, 'Etat actuel du diagnostic et du traitement des cancers du sein et de la langue', *Maroc-Médical*, January 1925(36), 1–14.

Tailhefer A. 1934, 'Traitement du cancer de sein', *La Médecine*, 15(8), 464–6.

Tailhefer A. and Dulac G. 1943, 'Traitement de l'epitheliome du sein par l'association de la chirurgie et de la roentgentherapie', *Bulletin de la Ligue Française Contre le Cancer*, December, 146–58.

Tailhefer A. 1954a, 'Essai sur le principe de therapeutique des cancers', *La Presse Médicale*, 4(16).

Tailhefer A. 1954b, 'A propos de la comunication de M. Denoix sur la diversité des cancers du sein. Principes actuels de la thérapeutique anti-cancereuse', *Memoires de l'Academie de Chirurgie de Paris*, 80, 565–8.

Tailhefer A. 1960, 'Néoplasmes mammaires et périmamaires', *Collection Médecine Interne*, fascicule 850, A-10.

Turesson I., Nyman J., Holmberg E. and Odén A. 1996, 'Prognostic Factors for Acute and Late Skin Reaction in Radiotherapy Patients', *International Journal of Radiation Oncology, Biology and Physics*, 36, 1065–75.

Vincent B. 1997, 'Genesis of the Pavillon Pasteur of the Institut du Radium of Paris', *History and Technology*, 13, 293–305.

Vincent B. 1999, 'Naissance et developement de la pratique therapeutique du radium en France, 1901–1914: Une substance entre médecine, physique et industrie', PhD thesis, Université de Paris VII.

Weisz G. 1998, 'A Moment of Synthesis: Medical Holism in France Between the Wars', in C. Lawrence and G. Weisz (eds) *Greater Than the Parts*, Oxford: Oxford University Press, 68–93.

Weisz G. 2005, *Divide and Conquer: A Comparative History of Medical Specialization, 1830–1950*, Oxford: Oxford University Press.

Wright J. 1985, 'The Development of the Frozen Section Technique, the Evolution of Surgical Biopsy and the Origins of Surgical Pathology', *Bulletin of the History of Medicine*, 59(3), 295–326.

Wolfromm G. 1957, 'Jean-Louis Roux Berger, 1880–1957', *La Presse Medicale*, 65, 26 October, 1745–6.

# 7

# Measured Responses: British Clinical Researchers and Therapies for Advanced Breast Cancer in the 1960s and 1970s

*Elizabeth Toon*

> ... [L]et us hope that, when all the differences have been quantified, there will be enough evidence for us to decide that some differences are important enough for one method to be preferred to another in some groups of patients. Alternatively, let us hope that the evidence will allow us to conclude that the results of some methods are so alike that it does not matter which we use.[1]
> – R. B. Welbourn, 1970

After the Second World War, newly developed systemic treatments for advanced breast cancer – endocrine therapy and cytotoxic chemotherapy – promised to improve and extend the lives of women with metastatic disease, while providing new insights about the nature of the disease itself. But clinician-researchers in Britain, North America and Europe quickly found that the benefits of these new interventions were unevenly distributed, and that the risks associated with them were substantial, sometimes even deadly. Given this unpredictable calculus of benefit and risk, these experts wondered how they could assess the utility of these therapeutic tools. Did endocrine therapy and cytotoxic chemotherapy work, they asked, for women with advanced breast cancer?

This essay considers how British clinician-researchers employed the tools of the new biomedicine – quantification, standardisation, and the clinical trial – to answer this question. British doctors, as I show, introduced, drew upon, and modified standardised measurement schemes for evaluating patients' responses to therapy. These schemes emerged locally, allowing clinician-researchers to understand patterns of response within a single institution's case series or trial, and these elite researchers then combined elements from different schemes to produce national standards for evaluating patient response in larger studies. Finally, the British rubric for assessing patient response would be integrated with that used by researchers from other countries, in order to create a standardised international framework for comparing patients' and study populations' responses, making it possible to consider the question of whether these therapies 'worked' definitively.

This story thus provides a clear example of the phenomenon Cambrosio *et al.* have recently termed 'regulatory objectivity'. Producing evidence in biomedicine, they argue, both requires and reproduces entities such as standards, markers, and protocols that coordinate collective action. Such shared entities allow their users to render judgements considered 'objective' even as they acknowledge that these standards, markers, protocols and the like are negotiated conventions, the dynamic products of consensus that enable the further production of consensus.[2] Certainly, British response criteria in advanced breast cancer developed much as Cambrosio *et al.* suggest, emerging first as local practices fitted to the priorities of individual research groups, which also provided their users with a reassuringly solid basis for difficult clinical judgements that required discriminating among confusing therapeutic possibilities. The criteria were then renegotiated for broader use by elite breast cancer researchers seeking to produce agreement about the value of particular therapeutic interventions, and eventually acquired a new life as general international standards for assessing therapeutic response in all solid tumour cancers.

But in tracing the development of these standards, we can also see how these clinician-researchers continued to grapple with the subjective considerations that arose out of their dual roles as researchers and clinicians, and that complicated their attempts to make objective data out of their patients' trajectories through illness and treatment.[3] Ilana Löwy's studies of leukaemia research and treatment remind us that making recourse to a collective ethos of decision-making about therapeutics can have an emotional function: standard criteria for assessing response both quantify and objectify the patient's trajectory, and that can allow clinician-researchers to reckon with difficult choices and come to terms with their own lack of control over their patients' conditions, by providing a seemingly objective, distanced calculus for directing treatment.[4] For the British clinician-researchers investigating and treating advanced breast cancer, standardised measurements of patient response were intended to provide that objectivity and distance. Indeed, these standards seemed to solve two problems at once, by showing the researcher whether a tested therapy acted on a cancer *and* by showing the clinician whether his/her patient was getting better.

And yet, the doctors who tried and trialled these new therapies found that despite their best efforts many responses blurred the line between objective and subjective, resisting measurement and comparability. As this essay shows, this posed a problem for the clinician-researcher dealing with advanced breast cancer: as a researcher, s/he could choose to measure only those responses which provided the 'cleanest' data and therefore apparently conclusive answers as to whether a therapy worked, but as a clinician, s/he might then find that such clean data failed to reflect the reality of the patient's trajectory, and thus give faulty guides for action. This conflict, between the roles of clinician and researcher, was one that was especially acute for these men and women, and for their contemporaries studying and treating other cancers: they were, after all, the first generation to fully adopt the clinical trial as a mode of evaluating potential interventions. But they were also the first to recognise which questions the trial could *not* answer. Standard response criteria, these clinician-researchers realised, might facilitate the comparison of therapeutic regimes, but did not really answer the question of whether a potentially beneficial but very risky therapy was 'worthwhile'. The 'worth' might differ considerably depending on whose 'while' was involved: that of the patient? the doctor? the health care system? Some of those who treated and studied breast cancer in the 1960s and 1970s would consider these questions; a few tried to come up with objective ways of answering them, while others shifted those questions aside, delegated them to other professionals, or reserved them for private discussion. Clinician-researchers, as we will see, could convert the confusing, contradictory, unreliable phenomena that constituted the subject-patient's trajectory through treatment into a definitive statement about whether a particular therapy worked. But collectively at least, they found it much more difficult to determine whether that therapy was worth it.

## The promise and problems of endocrine therapy

Before the Second World War, the doctors who treated advanced breast cancer had few therapeutic options available. The chief one was oophorectomy: Glasgow's George Beatson had announced in 1895 that surgical removal of the ovaries could provide some relief for women with 'inoperable' disease.[5] By the 1920s and 1930s, surgeons in North America, Britain, and Scandinavia often performed this ablative surgery on women with advanced disease, and by the 1940s, radiotherapists used X-rays to 'sterilise' the ovaries rather than removing them surgically.[6] But by the 1940s and early 1950s, the growing research interest in the breast cancer patient's 'hormonal environment' fused with increased technical ability to intervene in and manipulate that environment. Soon, clinicians at leading treatment centres had a growing array of potential tools available in the form of hormonal manipulation, and began using them in hopes of not only relieving pain, but also halting or even reversing disease progression.[7] Oestrogens, androgens, and other hormonal preparations, administered orally and by injection, were recognised as powerful agents for relieving symptoms in some patients. However, while some patients clearly benefited from hormone therapy, the consequences could be severe. Oestrogens frequently produced nausea, vomiting, water retention, and vaginal bleeding in the women who took them. Androgens, meanwhile, could produce a rapid and thorough response in some patients, enabling women previously bedridden by pain from bone and other lesions to walk, work, even marry. But this effect was not guaranteed, and many of those taking androgens seemed not to respond therapeutically but nonetheless experience such problematic effects as virilisation, hoarseness, hirsutism, hair loss, acne, weight gain, and increased libido.[8]

By the early 1950s, the availability of cortisone and the expanding technological resources of major clinical centres allowed doctors to alter the patient's 'hormonal environment' even more dramatically, by removing or disabling various glands. The two boldest of these interventions were adrenalectomy and hypophysectomy, the removal of the adrenal glands and the pituitary gland respectively, surgeries that were introduced in the 1950s and early 1960s.[9] Adrenalectomy was a relatively simple operation to perform and was sometimes performed at the same time as oophorectomy, but required that the patient be maintained for the rest of her life on cortisone. But despite its simplicity, adrenalectomy still carried a small but noticeable mortality rate, although by the middle 1970s it had been reduced to about one death for 62 operations. And slowing or temporarily halting disease progression was by no means assured: only about one-third of the patients receiving an adrenalectomy seemed to have regression of their disease as a result, and the proportion of 'responders' varied considerably from study to study. Furthermore, while some of those counted as 'responders' to adrenalectomy lived for several years after the procedure, the average survival time for 'responders' was said to be about 21 months, while 'non-responders' lived an average of nine months after the procedure.[10] By the early 1960s, those treating advanced disease with adrenalectomy began to wonder aloud about the ultimate value of the procedure, given the high proportion of apparent 'non-responders', left to live their remaining months with the consequences of adrenal removal and replacement.

Hypophysectomy presented the same quandary, but in even starker terms. When first introduced, surgical removal of the pituitary could only be done by neurosurgeons using a 'transfrontal' approach. Operative mortality was high in the first decade the procedure was performed, even though it was an intervention practiced only by very experienced surgeons in the very best centres. Potential consequences, gradually minimised by improving technique but never fully eliminated, included leaking of cerebrospinal fluid and damage to the brain and the optic nerves. And this was an extensive – and expensive – operation, requiring very skilled staff and considerable institutional support. In 1960, after describing a first substantial series of hypophysectomies, Guy's Hospital surgeon Hedley Atkins (later the president of the Royal College of Surgeons of England) wondered whether, even if it could be made to work in a substantial enough number of patients, hypophysectomy was a realistic intervention for advanced disease. After all, he pointed out, if some 7000 British women a year suffered from advanced breast cancer and thus might benefit from the procedure, how could the National Health Service possibly support this expensive, time-consuming intervention? He also suggested that because of shortages of facilities and personnel, there would be lengthy delays in offering hypophysectomy to women who might benefit from it, and those delays would in turn make the results attained by the treatment worse.[11]

The development of another technique for hypophysectomy, through the nose, made the procedure easier to achieve technically: now it could be performed by surgeons experienced in otolaryngology. But the consequences of the operation remained problematic: severe rhinorrhoea in some patients, meningitis in a few, and surgical mortality in a small but persistent number. Then other surgeons developed hypophysectomies that used radioactive materials to disable any portions of the pituitary that remained after surgery. After removing as much of the gland as possible, George Edelystn's Belfast team 'packed' the cavity with a wax containing radioactive yttrium oxide powder, and A. P. M. Forrest's group developed an implant containing radioactive yttrium, which could be 'screwed' into place.[12] Both of these techniques seemed to produce better results, and operative mortality from these procedures became rare. Nevertheless, patients still required long-term clinical attention and maintenance with cortisone and thyroid extract, many developed diabetes insipidus, and some suffered substantial damage to the optic nerves, resulting in blindness.[13] Studies comparing hypophysectomy and adrenalectomy further complicated matters. Hypophysectomy, if done well, produced a better and larger response, but if done badly – say, by a less experienced team – it meant certain disaster. Bilateral adrenalectomy could be more easily done, and could even be done at the same time as oophorectomy, but did not achieve quite such good results as hypophysectomy did.[14]

The willingness of cancer specialists to use these difficult, dangerous, but promising therapies was influenced by what seem to be powerful experiences with individual patients they had treated. In conference discussions and even in published papers, these clinician-researchers described women who were able to return to active lives, often for relatively lengthy periods of time. Some women's recoveries were remarkably impressive, leading investigators to use words like 'spectacular' and 'astonishing' (not terms usually employed by those treating advanced breast cancer) to describe them. Case descriptions told of bed-ridden women relying on opiates to manage pain, who underwent hormonal therapy and then returned normal activities:

> … a nullipara aged 51, who had been recurrence-free for 10 years after a radical mastectomy, when she developed backache, which persisted, increased in severity, eventually confining her to bed. Multiple deposits of secondary carcinoma filled most of the dorsal and lumbar vertebrae, and all the pelvic bones. Within a few days of the implantation of 500mg. testosterone the pain abated, walking was quickly resumed, and she is pain-free 14 months later.[15]

Given how profound and crippling many of these women's pain was, these rapid and thorough responses were no small achievement.

On the other hand, the effects of hormonal manipulation gave some investigators pause. Such therapies could induce profound bodily changes, from rapid and undesired weight gain to virilisation and changes in the sex drive. While clinicians worried about how their patients would respond to these changes, usually classing them as 'distressing' or 'very distressing', some also hinted – again, often in the less formal medium of conference discussions – that such changes, particularly in sex characteristics, disturbed them as well. For instance, one surgeon at a 1970 conference offered such a comment:

> I regard virilization as a very serious side-effect and I am wondering if it is ethically justifiable. Now I have seen cases of this sort where a woman's personality has been completely altered. She gets very

coarsened and her children are actually frightened when she approaches them. I think androgens ought to be withdrawn; but I do not think you can talk people out of it. ... Apart from libido, if you have to supply a straight razor with testosterone it is the best thing for a woman not to have treatment.[16]

Elite advocates of these therapies, who had more experience with finessing dosage and who dealt with a larger number (and variety) of patients, responded that lowered doses could produce substantial benefits and employed these therapeutic options accordingly, but how effective these arguments were beyond their peer group is not clear.

The increasing availability of cytotoxic and other chemotherapeutic preparations in the early 1960s only compounded the problematic calculus of therapeutic choice. Some clinician-researchers at leading British institutions began experimenting with cytotoxic preparations for their advanced breast cancer patients, but for most of them chemotherapy took a back seat to hormonal manipulation until at least the 1970s. Nevertheless, those who employed cytotoxic chemotherapy found that it presented the same unpredictable balance of problems and possibilities as administered hormones and endocrine ablation did. Most of the drugs used for chemotherapy entailed severe physical effects, especially nausea, vomiting, and hair loss. Surgeons and radiotherapists tried different combinations of chemotherapy, with regimens involving anywhere from two to five different drugs; infusing, injecting, and administering these was time-consuming and required patients to make long and frequent visits to the facilities where the therapy was delivered.

By the middle and late 1960s, then, those who treated women with advanced breast cancer had a large number of therapies available to them, each of which showed some promise of remission or slowed progression in some patients, but not in others. But there was still considerable debate over which of these to employ, when, on what kind of patients, and in what combination and order. Not surprisingly, leading clinical researchers took this opportunity to remind their colleagues how such a debate should be settled. In a typical statement, in this case addressed to the Royal College of Surgeons of Edinburgh, surgeon A. P. M. Forrest pointed out that in one mid-1960s survey of 211 patients, most had received at least three different types of treatments since they had developed recurrent disease, and a quarter of the patients had received five or more. 'Too often these patients are treated by a variety of endocrine or non-endocrine measures which depend more on the individual preference of the surgeon or radiotherapist than on scientific evidence of their relative merits', he opined.[17]

To find this evidence, Forrest and other elite clinician-researchers turned to the clinical trial, comparing different therapeutic options, and different combinations and sequences of therapeutic options for women with advanced breast cancer. These trials were intended not only to determine which therapies worked (and worked best), and in what combination and which order they should be delivered; they were also expected to help answer the question of *which patients* were likely to benefit from these therapeutic regimens. Given the serious consequences associated with both endocrine therapy and chemotherapy, and given that therapeutic success was apparently unevenly and unpredictably distributed, clinician-researchers hoped that by correlating the characteristics of patients who 'responded' to particular therapeutics, they might establish predictive indicators that would guide the future selection of patients likely to respond. Furthermore, this would not only save probable 'non-responders' from fruitless and frequently problematic interventions, but might even produce information about the nature of breast cancer as a systemic disease.

But these trials proved extremely difficult to conduct, and their results difficult to credit. First, trials in advanced breast cancer involved relatively small numbers of patients, as researchers needed subjects whose disease had recurred, but who were not yet too ill to withstand therapy itself. Second, the women who could be included in these trials had frequently already undergone some type of therapy that could potentially affect the outcomes of the new therapeutic regimen, possibly in an unknown way. For instance, some pre- or perimenopausal patients whose breast cancer had advanced had already had artificial menopause induced at the time that they received primary therapy, a practice known as 'prophylactic oophorectomy' or 'prophylactic castration' (also sometimes induced by irradiation rather than surgery). Given that many of the interventions being trialled for advanced disease were intended to manipulate the patient's hormonal environment, researchers asked, what difference did previous adjustments to that environment make? Third, women with advanced breast cancer were an incredibly varied lot, not just in terms of previous therapies received, but in terms of age, menopausal status, length of time between initial treatment and recurrence (known as the 'free period') and degree of illness, to mention but a few obvious categories. Researchers quickly agreed that trials needed to be stratified in order to account for this diversity, but this made the problem of small numbers more acute. Given all these complications, it is hardly surprising that most trialists enlisted expert statistical help to structure their protocols and analyse their results: it required no small amount of mathematical acrobatics to come up with a trial that would stand up to the scrutiny of colleagues and rivals, much less one that would produce statistically compelling evidence. But soon another problem trumped these: the question of response.

## What counts as a response?

The clinician-researchers who organised therapeutic trials in advanced breast cancer agreed that their goal was never to achieve a 'cure' of the disease, even if 'cure' was defined (as it was for primary treatment) as 'five-year-survival'. After all, the very definition of 'advanced' cancer was that the disease had spread or recurred, and that the patient's death would almost certainly be a result of the cancer. If the patient's eventual death from the disease was inevitable, what constituted therapeutic success for advanced breast cancer? It was clear that after therapy some patients benefited: they exhibited signs of disease regression, and they felt better, lived longer, and could do more of their normal physical activities. How, then, could this benefit be understood and compared, in a rigorous scientific manner?

The simplest measure of therapeutic benefit was length of survival after treatment, and this was commonly used. For instance, a combined case series of hypophysectomies found that from 12 to 46 per cent of the patients who received the operation were said to respond, and those 'responders' survived a mean 23 to 28 months after the procedure, whereas 'non-responders' survived an average six to ten months after the operation. But those were means, and the individual outcomes within both the responding and non-responding groups varied considerably.[18] Furthermore, average length of survival after treatment might look like an absolute number, but in practice only made sense when compared to survival after other therapies. But here, theories about the natural history of breast cancer complicated the situation. Perhaps, some specialists argued, some patients were naturally predisposed to live longer with their cancers, because even when disseminated they were less biologically aggressive, while a rapid decline might be preordained in other women with more biologically aggressive forms of the disease.[19] If that were true, the length of survival after treatment reflected not the effects of therapy, but the characteristics of underlying disease. Even changes of size in lesions might not be a sure indicator of a patient's progression or regression, some argued: perhaps it was the appearance of new lesions that really marked the progression of disease and thus the failure of treatment. Likewise, when a therapy seemed to have a negative effect on the patient's disease, that too might be due to the disease rather than the therapy. As one discussant at a 1970 workshop put it, 'How do you assess that treatment is making a patient worse when they may get worse anyway?'[20]

And it was not just the quantity of survival that was at issue: clinicians agreed that patients could have very different remissions qualitatively. For instance, one patient might have an immediate regression of disease, a long period with minimal symptoms, and then a short period of rapid decline before death; another patient could live the same amount of time, but with considerable illness and pain, no real remission, but just a slow, painful decline to death. Some researchers thus adopted two measures, intended to indicate both quantity and quality of remission: survival after treatment, and percentage of survival spent in remission.

That was just the start of the problem. As leading surgeons and others had eagerly adopted new measures like hypophysectomy and adrenalectomy, many in the late 1950s and early 1960s had reported the response rates obtained with these procedures – but, it quickly became apparent, without specifying what exactly they meant by 'response'. Did they mean that the disease had halted? Or did they mean that the disease had regressed? Did they mean symptomatic improvement? Did they mean increased survival times? Others reported the 'response' to treatment, but were careful to describe what, to them, counted as a response: take for instance this statement by two surgeons from Luton and Dunstable Hospitals trying combined adrenalectomy and oophorectomy: objective remission, they said, could be 'taken to include all cases where there was either regression or arrest of the disease after surgery as assessed by serial clinical, radiological, biochemical and histological measurements'.[21] (Specifically which measurements – and there were many from each category available – they never said.) But efforts to provide a definition of response hardly solved the problem, if every article had to contain such caveats. What counted as a response, anyway? And how could researchers be sure that their criteria of response made their results comparable to those of other researchers?

American clinical researchers addressed the problem first. Their early evaluations of endocrine therapy for advanced breast cancer were tabulated and published under the aegis of the American Medical Association's (AMA) Council on Pharmacy and Chemistry; these reports characterised such outcomes as 'subjective' response, regression in lesions of the soft tissue, lungs, and bones, and survival.[22] Initially, the precise parameters for measuring response seem to have been left up to the discretion of individual research groups. But by the early 1960s, groups like the Cooperative Breast Cancer Group (CBCG) and the AMA's Council on Drugs began to lay out in print apparently collective decisions about what should constitute a 'successful response' in American trials. These groups quickly made it clear that they were only interested in what they considered to be 'objective improvement': subjective responses were to be ignored in these reports, and only improvement – not arrest of disease or stationary lesions – was counted for evaluation purposes.[23] In the published protocol for evaluating thio-TEPA and HN2, two of the earliest cytotoxic compounds to be used in breast and lung cancer, malignant melanoma, and Hodgkin's disease, the Eastern Cooperative Cancer Chemotherapy Group did ask for assessments of 'patient reaction' (using a numerical scale), but spent far more time advising investigators as to how to take serial measurements of lesion size.[24] Eventually, the various American cooperative groups involved in the therapeutic evaluation of cytotoxic chemotherapy and hormonal therapy developed guidelines that would come to govern future assessment. The Cooperative Breast Cancer Group's standards were that

... there should be a measurable decrease of at least 50% of all demonstrable lesions, while the remainder are static. A successful response is inadmissible when there is progression of any lesion, appearance of new ones or absence of objective change in any metastatic lesion after 6 months of therapy.[25]

But these criteria defined 'successful response' as one that could be registered by measurable indicators – or rather, by easily and reliably measurable indicators. Changes in pleural effusions or in cerebral metastases were very difficult to measure; if a patient's cerebral metastases improved or pleural effusions decreased, she would likely feel and even demonstrate improvement, but the CBCG standards assumed that if the improvement in those indicators was truly significant, it would be reflected in other, more readily measurable indicators. Likewise, because it was very difficult to determine what changes in bone lesions meant, and observers frequently disagreed about what the source of those changes was, assessment of bone lesions remained a special case: if they were unchanged, that could be taken as evidence of remission.[26]

Furthermore, according to these criteria, only disease regression counted as a successful response – whereas halted disease growth and dissemination, whether temporary or relatively long term, did not count. So, as the American Albert Segaloff of the Ochsner Clinic explained at a conference on clinical evaluation in breast cancer, cases where a patient's disease failed to either regress *or* progress were counted as failures in US trials. Segaloff presented the reasoning behind this decision as follows: if many patients in a trial remained 'static' after receiving the drug or procedure, it was likely that many of the other patients would regress, and their regression would provide adequate numbers to prove that the intervention had 'worked'. In other words, the operative assumption was that if a therapy was likely to induce a static condition in some patients, it was certain to induce measurable improvement in others. By this logic, as Segaloff explained, 'Objective regression of the tumour is the sole criterion of therapeutic effectiveness'. Segaloff and his colleagues defined 'objective regression' fairly restrictively: all apparent tumours had to diminish in size, and more than half of the 'non-osseous' lesions had to decrease in size[27] or more than 50 per cent of the lesions had to improve while the rest remained static. If any new lesions appeared or any existing lesions progressed, the response was deemed a failure – even if other lesions improved significantly.[28] Apparently, the majority of the American groups consulted in determining these criteria of response agreed that 'a static situation is not evidence of therapeutic effect'.

British investigators, however, devised different solutions to the problem of response. J. L. Hayward, who worked with Atkins' Guy's Hospital group and with R. D. Bulbrook at the Imperial Cancer Research Fund (ICRF), was especially concerned with articulating definitions of response and with correlating response with patient characteristics in order to establish clearer indications for hormonal intervention.[29] Hayward and his colleagues combined two measures in order to determine whether a patient had 'responded' to treatment. The first was the Mean Clinical Value (MCV), initially developed by Arthur Walpole (best known for his later work with tamoxifen) and Edith Paterson in the late 1940s.[30] The MCV system, as modified by the Guy's breast clinic, involved initial recording of all lesions through drawing, photography, and X-ray. The status of the lesions was assessed every four weeks, and the patient X-rayed again on every third visit (thus, every 12 weeks). While the clinician assessed the lesions, he or she filled in a record card, recording a mark for each lesion: if it improved it scored a 2, if unchanged it received a 1, and if it progressed it was marked 0. If new lesions appeared, they were registered and given a mark of 0; if it was unclear whether a lesion had increased or decreased in size, the clinician could assign it a 1 and re-examine at the next monthly assessment. Then, for each body system, the clinician added up the marks given to lesions, and divided by the number of lesions, and then applied a multiplier of six for convenience. The trajectory for that system could then be followed, on a scale of 0 to 12; scores over six counted as regression of disease, and those under six counted as progression of disease. Then, the observing clinician added up the marks for all systems, and divided by the number of systems scrutinised, to come up with the Mean Clinical Value – a straightforward integer, the rise and fall of which could be plotted over time. To calculate the 'total MCV' for a patient, the investigator could calculate the area under the MCV curve on the patient's chart; additionally, he or she could calculate the average MCV overall for a particular intervention.[31] Hayward argued that this approach to assessing response had several advantages. First, it could be done on the spot: 'all decisions on the patient's response are made at the bedside'. Second, the investigator had regular indications of a patient's trajectory, rather than having to wait three or six months and then baldly assessing a treatment as success or failure. Finally, because the system tracked multiple lesions and organ systems, its proponents argued that it would 'provide a more accurate and representative picture of the results of therapy'.[32]

Neither the Guy's group nor the other leading British research group, the one led by A. P. M. Forrest at Cardiff and then at Edinburgh, were willing to simply classify responses as either success or failure. For instance, Forrest pointed out that such a binary classification was unrealistic for what his group saw in the clinic, so they allowed three categories for response to endocrine and other therapies for advanced breast cancer:

1) clear-cut objective regression in which there was a generalised reversal of all visible and radiological manifestations of the disease,
2) those without any response in whom the disease continued as it was before, and
3) an ill-defined middle group of 'equivocal remission' in whom one did not know whether the disease was responding or not.[33]

The Guy's group also used a tripartite classification for their 'success rate', with the categories being success, failure, and intermediate; 'intermediate' responses were then further subdivided into mixed responses, static responses, or temporary responses. Such a classification, they argued, more accurately revealed important trends.

At the several international conferences where these elite clinician-researchers discussed the issue of response in the 1960s and 1970s, another difference of opinion quickly emerged around how to monitor and validate results. All involved agreed that precision in measurement was important in assessing response – although they also bemoaned the technical problems of measurement, noting that all the seemingly 'objective' measurement systems they used were in fact susceptible to error and bias. A slight adjustment in how a patient sat for a photograph, for instance, could make a lesion appear to have changed; radiological assessments of bone changes (and the reasons for these changes) were notoriously variable. Thus, even what seemed to be objective measurement of objective response required a significant level of trust, as one researcher pointed out:

> Much of what we accept as evidence of response is based on the reliability, or at least on our thoughts as to the reliability, of the clinical investigator, because we know that it is perfectly possible to record certain changes which cannot be completely and objectively checked by other investigators. ... [When considering deep-seated lesions,] we must depend chiefly upon the honesty of the clinical investigator, and whether he will or will not accept the responses exactly as they do happen, rather than as he wishes they would happen.[34]

For these reasons, the elite clinician-researchers who devised the American classification of response argued that outside review – of X-ray films claiming to document changing lesions, for instance – was an essential part of the process of assessment: 'The ultimate decision regarding such response in any patient should follow impartial evaluation, and should rest with persons other than the investigator himself'.[35] Ideally, outsiders would review the records of each patient in a trial, and come to final agreement about what had happened to the patient. By contrast, Hayward of the Guy's group argued that the process of assessing MCV, done by the patient's clinician, was a more appropriate way to draw conclusions about a subject-patient's trajectory, claiming that 'the person who has attended the patient and known her throughout her illness would be the better judge of response than outside reviewers assessing retrospective data'.[36]

Nevertheless, within both of the respective breast cancer research communities, some investigators disagreed with the standards for assessing response put forth by their colleagues, while others reminded their colleagues that categorisation of responses could obscure as easily as it could reveal what was happening in a particular set of patients. M. J. Brennan, from Detroit's Henry Ford Hospital, articulated a growing unease about the binary, unambiguous categories embodied in the predominant American guidelines, when he pointed out that such response

criteria had grown out of those developed for systemic chemotherapeutic interventions for a very different set of diseases, the leukaemias and
lymphomas. When those cancers were treated with those therapies,
patient response was said to be easily apparent to investigators and
relatively unambiguous.[37] 'What', he asked, 'happens when one tries
to locate a concept of this special kind and experimental origin within
the context of a chronic disease like breast cancer?'[38] Statements about
breast cancer treatment, Brennan continued, must be 'approximate and
conditional':

> The attempt to categorise the spectrum of responses seen during the
> treatment of disseminated breast cancer into one category of regres-
> sions and another of therapeutic failures was an attempt to make
> breast cancer treatment as suitable a subject for statistical methods
> as the treatment of acute leukemia.[39]

Brennan further suggested that bald characterisations of response 'worked'
if an investigator wanted to compare therapeutic regimens – but they did
not 'work' if a clinician wanted to understand the reality of advanced
breast cancer and the merits of the various therapies available for it for
particular patients. If investigators could bring themselves to admit this,
he continued, the question became 'How can we express the results
of chemotherapeutic treatment of disseminated breast cancer without
making arbitrary decisions, creating artificial categories or inflating the
significance of our efforts in the minds of our colleagues?'[40] Likewise,
the ICRF biochemist R. D. Bulbrook reminded his fellows at a 1974
conference that

> Response to treatment is a continuous variable, not an all-or-none
> phenomenon and, like survival, is log-normally distributed. This
> adds to the difficulties in that most workers prefer to express their
> results as 'success' or 'failure', which means the introduction of an
> arbitrary cut-off point varying from worker to worker ... If enough
> variables are measured, in a small enough series, a perfect separation
> of responders and non-responders can usually be achieved. But the
> same degree of separation could probably be achieved by measuring
> variables such as shoe and hat size.[41]

If researchers wanted to use responses to therapy as a way to reveal prognostic factors and new knowledge about breast cancer's natural history,
then playing with the numbers and with arbitrary classifications would

defeat that purpose, revealing instead putative 'prognostic factors' that were far from real.

After several conferences and workshops in the late 1960s and early 1970s where the issue of response was discussed, in 1974 the newly formed British Breast Group decided to create its own standards for assessment of response.[42] A committee chaired by A. P. M. Forrest came up with a system which in one important respect resembled that of the American cooperative groups: the committee endorsed external review of data that was difficult to categorise. But the system for assessing response set up by the British Breast Group (BBG) committee did reflect the earlier preference amongst British investigators for a system that incorporated some flexibility and ambiguity.[43] While the categories of regression and progression were fairly clear-cut, a third category – variously called 'intermediate', 'equivocal', or 'partial' response – was established, and the investigator could further classify intermediate responses as 'mixed', 'static', or 'temporary'. Some investigators then took advantage of this third category when considering the strength of the others, by presenting results for 'definite successes' (excluding both definite failures and intermediate responses) or 'definite failures' (excluding both definite successes and intermediate responses).

The flexibility and realism of the BBG system, its creators agreed, was necessary, to uncover the 'real' commonalities among those women who responded to therapy. Obviously, this knowledge would be immediately beneficial, as it would allow clinicians to predict who was likely to respond to therapy – and even be able to offer more certain prognostication about the future of their patients. Equally important, though was the improved understanding of breast cancer itself that seemed likely to emerge: if manipulating the hormonal system of the woman with advanced breast cancer affected the course of her disease, that meant that the endocrine system was likely to be fundamentally implicated somehow in breast cancer's natural history, and could potentially then offer the tools for a systemic attack on the disease. Cancer researchers had mulled over this possibility since Beatson's oophorectomies initially suggested it, and by the post-war decades, increasingly sophisticated assay methods facilitated the study of endocrine function and breast and prostate cancer. W. T. Irvine, a surgeon at St Mary's (London), directed one group investigating that relationship, by tracking the urinary excretion of hormones in women undergoing endocrine procedures, and the accumulation of oestrogens in tumour tissue.[44] At Guy's Hospital, Atkins' group began pursuing this question in the 1950s; Atkins hired Hayward, who then began his collaboration with Bulbrook, producing a joint

Guy's/ICRF research programme that pivoted on measuring the urinary excretion of hormones and their metabolites. They not only attempted to correlate these measurements with patients' responses to endocrine manipulation, but also ran a predictive study on the island of Guernsey, collecting and analysing urine from thousands of women for several years, in hopes of discovering a correlation between hormonal status and the later development of breast cancer.[45] Unfortunately, creating even a predictive indicator of how women might respond to endocrine therapy based on measurements derived from urinary analysis proved exceptionally difficult. In the end, Hayward and Bulbrook came up with a 'discriminant function' that drew together several measurements from urinary analysis to produce what seemed to be a good prognostic index for further hormonal manipulation in individual patients.[46] But this biochemical analysis and the discriminant function proved difficult to use, and many other groups – especially outside Britain – soon abandoned it. Forrest's group in Edinburgh pursued similar investigations, as did several teams in North America.

Shortly after the BBG had established its system, key breast cancer researchers in the UK, the US, and elsewhere decided it was necessary to build an international system that would further standardise assessment of response for advanced breast cancer, so that studies would be internationally comparable. In 1975, an International Union Against Cancer/ Union International Contre le Cancer (UICC) committee was created, with Hayward at its head and including surgeons, radiotherapists, and medical oncologists from Continental Europe, the US, and Japan. The system they eventually produced was a hybrid of the American and British approaches, but more closely resembled the systems used by the American cooperative oncology groups.[47] It offered detailed instructions for measuring lesions and the UICC committee suggested the use of many new technologies for doing so. A representative number of measurable lesions were to be monitored, and 'regression' was deemed to have happened when all lesions disappeared, or when the sum of the products of the measurements of monitored lesions had decreased by 50 per cent and no new lesions had appeared. But the system also had four categories: complete response, partial response, no change, and progressive disease. To qualify for 'partial response', not every lesion had to improve, but none could get worse. (A year later, in 1978, the criteria were amended to be stricter: a partial response now required more than a 50 per cent decrease in measurable lesions, demonstrable improvement in non-measurable lesions, and no new lesions.[48]) The UICC's guidelines also suggested that investigators stratify results by patients' menstrual status, into three groups:

pre-menopausal, early post-menopausal (from one to five years after the last period), and late post-menopausal (more than five years after the last period). Finally, the UICC committee strongly encouraged investigators to adopt some measurement systems that would integrate alternative senses of 'response'. In particular, they suggested that researchers also assess 'performance' in the base-line patient assessment and after treatment, using the Karnofsky Scale.[49]

In 1977, the Lancet heralded the UICC's new guidelines, hoping that the adoption of such standards would 'allow comparisons between different centres, and which are not bedevilled by the variations which can arise from differing optimisms and enthusiasms of the investigators'. But they concluded by noting that additional, standardised measurements of response were still required – for the aspects of response that remained difficult to quantify: 'Much harder to evaluate, but of equal importance, is symptomatic improvement and performance. The next task for the cancer groups is to provide guidelines for assessment of these aspects of response to treatment'.[50] But would that task prove amenable to collective decision-making?

## Subjective and 'worthwhile' responses

As we saw at the beginning of this essay, the unintended consequences of new therapies for advanced breast cancer were widely acknowledged to impose a significant burden on the women subjected to them. Ablative procedures like adrenalectomy and hypophysectomy required lifetime replacement therapy and had significant side-effects; unintended consequences of these therapies, such as nerve damage, meningitis, and even death, remained a real, if decreasing problem. Administered hormones and cytotoxic drugs brought on sometimes dramatic bodily changes, including severe nausea, bleeding, hair loss, and virilisation. Even when these therapies seemed to 'work', producing an improvement in symptoms and pain control, clinician-researchers asked, did the therapeutic gain outweigh the burden placed on patients by the therapy? If one came up with a way to assess 'response', could one also determine whether a response was 'worthwhile'? To do so was vital, the men and women developing, testing, and applying these therapies argued, but they agreed such determinations were difficult to make.

First, it was clear that subjective responses were just that: subjective. All those reporting on their use of endocrine and cytotoxic therapies noted that patients reported subjective response – even when 'objective' measures of response showed no improvement. Patients demonstrating objective response almost always reported subjective response, but many patients with no objective response reported that, despite the absence of measurements proving tumour regression and lesion healing, they *felt* better and had regained physical function. Given that the action of some of these therapies (and aspects of breast cancer's progression itself) was not fully understood, could this subjective improvement be ignored, even if objective evidence of improvement remained elusive? The French investigator Phillippe Juret, for instance, explained that in his series of patients receiving a radioactive yttrium implant, only about one-third improved 'objectively' – but 86 per cent reported that they felt better.[51] Was such a procedure (which required significant resources and continued monitoring of the patient) worthwhile?

Indeed, many clinician-researchers, in a philosophical mood, used this very word *worthwhile* when they asked deeper questions about the value of treatment. But they were not always clear about whose 'while' a procedure was 'worth' – sometimes they claimed to be speaking for the patient, anxious to spare her time and dignity, but it could also be asked whether it was worth their own time and energy to perform such procedures. (Hedley Atkins' musings, discussed above, on whether Britain's health infrastructure could handle 7000 hypophysectomies a year is one of the few outright statements of this type, though.) Who could, who should, judge the value of such complicated trade-offs? Some investigators implied that they could make that judgement, on both caring and practical grounds. W. T. Irvine, for instance, argued that

> Too much has been written about the benefits of remission.... All too often within 6–8 months there was evidence of fresh progress either at the old sites or in new metastatic deposits. Any humane clinicians treating a large series of these patients have their beds inundated with such relapsing patients, to say nothing of those who fail to respond and remain hospital patients to the end.[52]

A few years later, he reiterated this argument, and although he believed that the procedures involved in adrenalectomy and hypophysectomy had improved, he remained doubtful that they were worthwhile, concluding that

> There is something unconvincing about the advocacy of a major surgical regime whose total achievement is augmentation of mean

survival by a few months in less than one-third of those undergoing operation.[53]

But R. B. Welbourn, who was quoted at the opening of this paper, volunteered a different opinion: he had treated a nurse's advanced breast cancer with hypophysectomy, and she had gone blind as a result. But she had also been relieved of horrible bone pain for the remaining nine months of her life. She told him it was well worth the blindness to be relieved of her pain, but how, Welbourn wondered aloud in a conference discussion, could this be reflected in the published literature appraising response to treatment?[54]

Implicit in such comments is the realisation that, while surgeons and others tried to remain aware which consequences would be particularly 'distressing' to their patients, patients differed markedly in terms of what they considered distressing, and what they considered worth the distress. Some male clinicians considered that this might be a gender barrier, and debated whether a (presumably female) social worker should be enlisted to help appraise the patient's subjective response. Others worried about introducing yet another player into the therapeutic relationship, and instead suggested asking 'female medical graduates' on their team to enquire.[55] Interestingly, though, none discussed how their quest to make sense of responses would have reshaped the experience of the advanced breast cancer patient who was, almost certainly unknowingly, part of one of these trials. By virtue of the trialist's need for good measurements, these women would have been subjected to: extensive base-line assessment and frequent follow-up appointments; vaginal smears and urine assays to determine hormone levels; repeated photographic, X-ray, and palpatory assessment of lesions and later full bone and other scans; and in some cases very complicated treatment regimens. But even if clinician-researchers were unsure whether a particular therapy was worthwhile, objectively or subjectively, these experts seemed to take it as given that putting their subject-patients through extensive and intimate procedures to *evaluate* a therapy was unquestionably worthwhile.

In the end, although these researchers commented on, and clearly worried about the question of subjective response, their chief tactic was to try to fit it into the same kinds of parameters as they had fit 'objective' response. By the early 1970s, investigators increasingly assessed patients' performance after therapy and tried to gather data on patient response to therapy. The Karnofsky Scale for performance, for instance, offered a means for conveying the general state of a patient after her therapy, to go with the more precise measurements of her lesions. T. J. Priestman and Michael Baum borrowed the linear analogue self-assessment (LASA) from psychologists. Here, patients could assess their own appreciation of their illness by marking how they felt on a scale, and could also indicate whether they thought treatment was helping; Priestman and Baum also had the LASA administered both in the presence of doctors and when the patient was alone, to see if the doctor's presence might create discrepancies. The LASA, Priestman and Baum admitted, had significant problems, but still gave some indication of subjective response:

> The whole L.A.S.A. system may be criticised for only giving a superficial view of an individual's emotions and difficulties. However, we had no intention of making an in-depth psychological study of individual patients. What we needed was a readily comprehensible, convenient, and reliable way for patients to make their own assessments as to their quality of life during and after treatment.[56]

Efforts like these allowed those researching and treating advanced breast cancer to acknowledge the less quantifiable but still important effects of new treatments for patients undergoing a particular new therapy. But even decades later, some critics argued that these efforts to make sense of subjective response, and what a cancer treatment might mean to the people receiving it, remained seriously flawed. Physicians and surgeons, one study argued, 'could as not adequately measure ... patients' quality of life', because even on standardised scales, different doctors' scores showed wide variability.[57] And that was only if the doctors involved bothered to assess patients' quality of life. In one study conducted by social researchers from the Royal Marsden, of the clinician-researchers they surveyed, only one-third bothered to routinely measure quality of life when conducting clinical trials – and only one-fifth of the surgeons did. This was, the study's authors pointed out, because many of these elite researchers felt that 'QoL [quality of life, now reduced to an acronym] can be assessed adequately without formal instruments' or that it was simply too time-consuming.[58] It is no surprise that one author of this study was Cicely Saunders, the palliative care pioneer. For those promoting palliative care as a necessary element of cancer care, understanding quality of life and other responses that defied objectification was one of their chief concerns.[59]

## Conclusion

What, then, does the story of the evolution, negotiation, and renegotiation of standardised response criteria for advanced breast cancer therapy tell us? Most obviously, it helps illustrate some fundamental differences between British and US approaches to research on, treatment of, and theorising about breast cancer. More tolerant of intermediate categories and more interested in the promise of hormonal therapy, British researchers tended to see response as a continuous but classifiable phenomenon, whereas most of their American, chemotherapy-oriented colleagues endorsed 'all-or-nothing' criteria that resolved the question 'does this therapy work?' with a clear yes or no answer. Both groups were concerned to objectify response, to create systems of measurement that would make the assessment of outcomes comparable, even if they were willing to admit – in informal circumstances at least – that their systems of measurement might introduce rather than reveal difference. In everyday treatment surroundings and in informal professional settings, British clinician-researchers still considered what subjective response could tell them and what it meant for their patients. But by the 1970s, as the institutional case series was finally squeezed out of the pages of medical journals by statistical analyses of large trials, it became easier to keep those discussions out of formal medical discussion venues, which instead chronicled aggregate success and failure, statistically evaluated, in multi-centre clinical investigation. Speculation, philosophical musing, and subjective response were matters for letters to the editor and informal discussion. I suspect that American clinician-researchers also saved their discussion of subjective response for other, informal venues.

Meanwhile, in some ways, the problem of response solved itself, thanks to two developments in the 1970s and 1980s: the introduction of tamoxifen as a treatment for advanced breast cancer in the 1970s (and later, as a treatment for primary breast cancer), and the introduction of hormone receptor assays to help clinicians predict response to treatment. Since tamoxifen was much more easily tolerated by patients than any of the other endocrine treatments, its therapeutic calculus was far less fraught. This drug could be easily administered and stopped, without the extremely serious risks and very problematic side-effects – some of which persisted for the remainder of the patient's life – that came along with administered hormones, surgical endocrine ablation, and cytotoxic chemotherapy.[60] Then, the elaboration of the hormone receptor concept in breast cancer treatment and the introduction of hormone receptor assays offered an approach to predicting response that appeared more viable and standardisable than analyses of hormone excretion patterns had been able to manage.[61] This meant it was easier for medics to select patients expected to respond to therapy, and allowed them to try to spare the women unlikely to respond to the physical and emotional trauma of ineffective, potentially harmful therapy.

As the tools for measuring and predicting response improved, the response criteria originally negotiated and codified for advanced breast cancer acquired a new and more powerful life, when the UICC guidelines for assessing response became the basis for response criteria in cancer studies generally. In the late 1970s, the World Health Organization (WHO) brought together representatives from research organisations around the world to discuss the assessment of therapeutic response; their conclusions were issued as a WHO technical report, and published in the world's leading cancer journals. Nevertheless, definitions and criteria for measuring drug toxicity took up as many pages as did definitions and criteria for measuring drug response – perhaps in response to the concerns of the pharmaceutical firms and official regulatory agencies now involved in the reporting of trials on new therapies?[62] Subjective response, by contrast, got a single short paragraph:

> Definition of subjective response is difficult because there are so many factors that can influence it. Despite this difficulty a response (e.g. weight gain or decrease in pain) can nevertheless be of great importance to the patient and may alert the physician to the possibility of an objective response.[63]

Subjective response was thus acknowledged in the research context, but that acknowledgement was primarily in relation to the 'real' objective response – not as a phenomenon in its own right and deserving of the researcher's attention.

The importance of this story, then, lies in the way that the questions around response exposed the divergent, even conflicting concerns of clinicians and researchers, at a time when cancer treatment was thought to be best conducted by practitioners who were simultaneously clinicians *and* researchers. As we have seen, for the men and women who conducted trials of treatments for advanced breast cancer, the goal was (as it always is in clinical trials) to transform the vagaries of individual treatment into objective data points that transcended the anecdotal, which in turn would provide an expert consensus about the 'best' path of action. Today this process has evolved into what we now call evidence-based medicine,[64] where the presumption is that to count as 'evidence', information must be objective. As elite clinician-researchers of the 1960s and 1970s investigated these risky therapies, and considered what the results from these therapies could reveal about breast cancer, they sought to transform the messiness of experience into, as much as was possible, objective data, or at least data that seemed (as Cambrosio *et al.*'s model of 'regulatory objectivity'[65] would suggest) objective enough for their purposes. Any messiness and variability that remained was labelled subjective response, a close cousin to other fuzzy phenomena like 'quality of life'.

As we saw, these experts did not dismiss the philosophical and practical problems raised by subjective response. As researchers, they considered how it complicated their attempts to answer scientific questions, and as clinicians, they admitted that subjective response made understanding the patient's personal, individual trajectory through illness and treatment more difficult. Creating standardised rules for measuring response and determining that only some responses merited measurement solved the problems that faced them as researchers; but the quiet questioning of these standards suggests that as clinicians, the men and women treating advanced breast cancer patients remained somewhat unsatisfied with this negotiated objectivity. Nevertheless, once they had determined as researchers that subjective response was not suitable for reporting in trials, they left it for informal discussion, or wrote separate articles about quality of life and subjective responses, or left it for analysis by other types of professionals, such as social scientists, nurses, and palliative care experts. In other words, by moving discussions of subjective response to other venues, clinician-researchers effectively removed this persistent messiness and variability from the discussions of new therapies that counted most in biomedicine, the analyses of big multi-centre trials in medical journals. In so doing, they also removed it from the discussions that provide the 'evidence base' for evidence-based medicine today.

## Acknowledgements

The research for this essay was done as part of the Constructing Cancers programme grant, sponsored by the Wellcome Trust. I thank the organisers of the 'Institutions of Objectivity' workshop held at McGill University, April 2007 for the opportunity to present an earlier version of this paper, and to the participants at that conference for their questions and feedback. My thanks also to Carsten Timmermann, John Pickstone, Jennifer Gunn, and Lyn Schumaker for their critiques and suggestions.

## Notes

1 Welbourn R. 1970, p. 49.
2 Cambrosio *et al.* 2006; Cambrosio *et al.* 2009.
3 See Marks 1997, especially Chapter 2, for a discussion of the emergence of therapeutic researchers; on the clinical trial in post-war Britain, see for instance Valier and Timmermann 2008.
4 Löwy 1995; Löwy 1996. See also Christakis 1999 for a thoughtful sociological analysis of how doctors manage uncertainty, their professional roles, and their emotional selves in prognostication.
5 The chief recent histories of twentieth-century breast cancer treatment focus on the treatment of primary disease, but do discuss Beatson's work and the use of oophorectomy in that context; see Aronowitz 2007 and Lerner 2001.
6 See for instance Anonymous 1939 and Paterson 1936.
7 Surprisingly little has been written about the role of hormone treatment in breast cancer, either by scholars interested in the history of cancer or those concerned with the history of hormones, although there is a good brief overview of early developments in Sengoopta 2006, pp. 195–200. The chief exception is Austoker 1988, pp. 215–21 and 244–54, which discusses the biochemical and clinical collaboration of the Imperial Cancer Research Fund's R. D. Bulbrook and Guy's Hospital's John L. Hayward; see also Hayward 1970 for a comprehensive discussion of that research. For an excellent technical and historical overview, which highlights both animal-based laboratory research and clinical work, see Howell *et al.* 1997. For general information about clinical endocrinology at mid-century, see Welbourn 1993, especially pp. 490–4 and 505–6. On the synthesis of steroids, including cortisone, see Gaudillière 2005, Quirke 2005, Slater 2000, and Marks 1992; also Watkins 2007, pp. 24–6 on Dodds and the development of synthetic oestrogen.
8 For an overview of hormonal approaches to advanced breast cancer generally, see Heuson 1974, p. 135. On oestrogens, see Haddow *et al.* 1944; Anonymous 1944; Walpole and Paterson 1949. On androgens, see Galton 1950. Finally, Yolanda Eraso (2010) has examined in more detail how gender conventions shaped medical concerns about the virilising effects of hormonal therapy for breast cancer, comparing how American, British, and Argentinian doctors viewed these unwanted consequences of therapy.
9 On adrenalectomy, see Huggins and Bergenstal 1952; on hypophysectomy, see especially Luft and Olivecrona 1957.
10 Burn 1974, p. 99.
11 Atkins *et al.* 1960, p. 1153.
12 Edelstyn *et al.* 1958 and Edelstyn *et al.* 1968; Stewart *et al.* 1969.
13 Burn 1974, p. 102.
14 Atkins in the introduction to his (ed.) 1974, p. 5.
15 Galton 1950, p. 41.
16 See the discussion in Joslin and Gleave (eds) 1970, p. 38. The female speaker who had presented these results, Dr Vera Jones, responded that lower doses engendered fewer changes, and (perhaps drily) pointed out that there were other ways for women to manage facial hair besides safety razors.
17 Forrest 1967, pp. 192–4.
18 Burn 1974, pp. 102–3.

19  For a discussion of these concerns about biological variation and aggressive cancers in the US context, see Lerner 2001, pp. 92–114.
20  T. Hamilton in Joslin and Gleave (eds) 1970, p. 83.
21  Barlow and Meggitt 1968, p. 811.
22  Council on Pharmacy and Chemistry 1949; Council on Pharmacy and Chemistry 1951.
23  Subcommittee on Breast and Genital Cancer 1960.
24  Zubrod *et al.* 1960.
25  Heuson 1974, p. 115.
26  Heuson 1974, pp. 114–16.
27  Again, because measuring bone lesions was so problematic, they were allowed to remain static under this definition of 'objective regression'.
28  Segaloff 1966, pp. 126–9.
29  Austoker 1988, pp. 244–51.
30  Walpole and Paterson 1949, p. 783.
31  Hayward 1966, pp. 134–7.
32  Hayward 1966, p. 137.
33  A. P. M. Forrest in Forrest and Kunkler (eds) 1968, p. 220.
34  Escher 1966, p. 6.
35  Cooperative Breast Cancer Group 1964, pp. 1069–72.
36  Hayward in discussion following Brennan 1966, p. 172.
37  Whether response to chemotherapy in leukaemia and lymphoma *was* as apparent and unambiguous as Brennan claimed is of course an entirely different question.
38  Brennan 1968, pp. 197–8.
39  Brennan 1968, p. 199.
40  Brennan 1968, p. 201.
41  Bulbrook 1974, pp. 181, 182–3.
42  Anonymous 1974, p. 60.
43  British Breast Group 1974, pp. 38–9.
44  Folca *et al.* 1961, pp. 796–8; Irvine *et al.* 1961. On the important contribution of Irvine's team, see Howell *et al.* 1997, p. 5.
45  Austoker 1988, pp. 248–50.
46  Bulbrook and Hayward 1965; Hayward and Bulbrook 1965.
47  Anonymous 1977a; Anonymous 1977b.
48  Anonymous 1978.
49  On the development of the Karnofsky Scale and other modes of assessing patient status and performance, see Timmermann 2011.
50  Anonymous 1977c, p. 840.
51  Juret 1966, p. 189.
52  Irvine *et al.* 1961, p. 795.
53  Irvine 1964, pp. 161–7.
54  R. B. Welbourn in Joslin and Gleave (eds) 1970, p. 89.
55  Eric N. Gleave in Joslin and Gleave (eds) 1970, p. 89.
56  Priestman and Baum 1976, p. 900.
57  Slevin *et al.* 1988.
58  Saunders and Fallowfield 1996.
59  On the development of palliative care, see Lewis 2007, esp. pp. 123–33.
60  For the introduction of tamoxifen for advanced breast cancer, see Cole *et al.* 1971.

61  On hormone receptors and the prediction of response to therapy, see Howell *et al.* 1997, pp. 5–7; also Jensen *et al.* 1971 and McGuire *et al.* 1975. The introduction and standardisation of receptor assays raised many of the same questions as response criteria had – for instance, whether or not hormone receptor status was binary or continuous, and how diverse local and national approaches to measurement could be reconciled: see for instance King *et al.* 1979, King and Roberts 1979, and the other essays in King (ed.) 1979 for British approaches to these questions.

62  See for instance the nine page appendix of toxicity criteria produced by the Southwest Oncology Group in Green and Weiss 1992.

63  Miller *et al.* 1981, pp. 207–14.

64  Daly 2005.

65  Cambrosio *et al.* 2009; Cambrosio *et al.* 2006.

## Works Cited

Anonymous 1939, 'Ovarian Irradiation in Incurable Breast Cancer', *Lancet*, 233, 6033, 884–5.

Anonymous 1944, 'Stilboestrol for Advanced Breast Cancer: A Combined Investigation [Reports of Societies]', *British Medical Journal*, ii, 4356, 20–1.

Anonymous 1974, 'British Breast Group [Notes and News]', *Lancet*, 304, 7871, 60.

Anonymous 1977a, 'Assessment of Response to Therapy in Advanced Breast Cancer: A Project of the Programme on Clinical Oncology of the International Union against Cancer', *British Journal of Cancer*, 35, 292–8.

Anonymous 1977b, 'Assessment of Response to Therapy in Advanced Breast Cancer: A Project of the Programme on Clinical Oncology of the International Union against Cancer', *Cancer*, 39, 1289–94.

Anonymous 1977c, 'Advanced Breast Cancer', *Lancet*, 309, 8016, 840.

Anonymous 1978, 'Assessment of Response to Therapy in Advanced Breast Cancer (An Amendment)', *British Journal of Cancer*, 38, 201.

Aronowitz R. 2007, *Unnatural History: Breast Cancer and American Society*, Cambridge: Cambridge University Press.

Atkins H., Falconer M. A., Hayward J. L. and MacLean K. S. 1960, 'Adrenalectomy and Hypophysectomy for Advanced Cancer of the Breast', *Lancet*, 275, 7135, 1148–53.

Atkins H. (ed.) 1974, *The Treatment of Breast Cancer*, Lancaster: Medical and Technical Publishing.

Austoker J. 1988, *A History of the Imperial Cancer Research Fund, 1902–1986*, Oxford: Oxford University Press.

Barlow D. and Meggitt B. 1968, 'Clinical Indices to the Response Rate of Advanced Breast Cancer to Bilateral Adrenalectomy and Oophorectomy', *British Journal of Surgery*, 55, 809–16.

Brennan M. 1968, 'The Value of Hormonal and Chemotherapeutic Treatment in Disseminated Breast Cancer', in A. Forrest and P. Kunkler (eds) *Prognostic Factors in Breast Cancer: Proceedings of First Tenovus Symposium, Cardiff 1967*, Edinburgh and London: E & S Livingstone, 197–210.

Brennan M. 1966, 'Indices of Response to Breast Cancer Therapy', in J. Hayward and R. Bulbrook (eds) *Clinical Evaluation of Breast Cancer: Proceedings of the*

*First Imperial Cancer Research Fund Symposium, 1965*, London and New York: Academic Press, 141–68.

British Breast Group 1974, 'Assessment of Response to Treatment in Advanced Breast Cancer', *Lancet*, 304, 7871, 38–9.

Bulbrook R. 1974, 'Tests of Prediction', in H. Atkins (ed.) 1974, *The Treatment of Breast Cancer*, Lancaster: Medical and Technical Publishing, 177–216.

Bulbrook R. and Hayward J. 1965, 'The Possibility of Predicting the Response of Patients with Early Breast Cancer to Subsequent Endocrine Ablation', *Cancer Research*, 25, 1135–9.

Burn J. 1974, 'Endocrine Therapy – Ablative Surgery', in H. Atkins (ed.) 1974, *The Treatment of Breast Cancer*, Lancaster: Medical and Technical Publishing, 87–112.

Cambrosio A., Keating P., Schlich T. and Weisz G. 2006, 'Regulatory Objectivity and the Generation and Management of Evidence in Medicine', *Social Science and Medicine*, 63, 189–99.

Cambrosio A., Keating P., Schlich T. and Weisz G. 2009, 'Biomedical Conventions and Regulatory Objectivity: A Few Introductory Remarks', *Social Studies of Science*, 39, 651–64.

Christakis N. 1999, *Death Foretold: Prophecy and Prognosis in Medical Care*, Chicago: University of Chicago Press.

Cole M. P., Jones C. T. A. and Todd I. D. H. 1971, 'A New Anti-Oestrogenic Agent in Late Breast Cancer: An Early Clinical Appraisal of ICI 46474', *British Journal of Cancer*, 25, 270–5.

Cooperative Breast Cancer Group 1964, 'Testosterone Propianate Therapy in Breast Cancer', *Journal of the American Medical Association*, 188, 1069–72.

Council on Pharmacy and Chemistry, American Medical Association 1949, 'Estrogens and Androgens in Mammary Cancer: A Progress Report', *Journal of the American Medical Association*, 140, 1214–16.

Council on Pharmacy and Chemistry, American Medical Association 1951, 'Current Status of Hormone Therapy of Advanced Mammary Cancer', *Journal of the American Medical Association*, 146, 471–7.

Daly J. 2005, *Evidence-Based Medicine and the Search for a Science of Clinical Care*, Berkeley: University of California Press.

Edelstyn G., Gleadhill C. and Lyons A. 1968, 'Total Hypophysectomy for Advanced Breast Cancer', *Clinical Radiology*, 19, 426–32.

Edelstyn G., Gleadhill C. A., Lyons A. R., Rodgers H. W., Taylor A. R. and Welbourn R. B. 1958, 'Hypophysectomy Combined with Intrasellar Irradiation with Yttrium-90', *Lancet*, 271, 7018, 462–3.

Eraso Y. 2010, 'Gendering Breast Cancer Treatments in International Perspective', paper given at 'Politics and Practices: The History of Post-War Women's Health' conference (23 October), University of Manchester, Manchester UK.

Escher G. 1966, 'Methods of Measurement of Soft Tissue Lesions', in J. Hayward and R. Bulbrook (eds) *Clinical Evaluation of Breast Cancer: Proceedings of the First Imperial Cancer Research Fund Symposium, 1965*, London and New York: Academic Press, 1–5.

Folca P. J., Glascock R. F. and Irvine W. T. 1961, 'Studies with Tritium-Labelled Hexoestrol in Advanced Breast Cancer: Comparison of Tissue Accumulation of Hexoestrol with Response to Bilateral Adrenalectomy and Oophorectomy', *Lancet*, 278, 7206, 796–8.

Forrest A. 1967, 'Clinical Studies in Advanced Breast Cancer', *Journal of the Royal College of Surgeons of Edinburgh*, 12, 192–206.

Forrest A. and Kunkler P. (eds) 1968, *Prognostic Factors in Breast Cancer: Proceedings of First Tenovus Symposium, Cardiff 1967*, Edinburgh and London: E & S Livingstone.

Galton D. 1950, 'Androgen Therapy in 70 Cases of Advanced Mammary Carcinoma', *British Journal of Cancer*, 4, 20–58.

Gaudillière J.-P. 2005, 'Better Prepared than Synthesized: Adolf Butenandt, Schering Ag and the Transformation of Sex Steroids into Drugs (1930–1946)', *Studies in History and Philosophy of Biological and Biomedical Sciences*, 36, 612–44.

Green S. and Weiss G. 1992, 'Southwest Oncology Group Standard Response Criteria, Endpoint Definitions and Toxicity Criteria', *Investigational New Drugs*, 10, 239–53.

Haddow A., Watkinson J. M., Paterson E. and Koller P. C. 1944, 'Influence of Synthetic Oestrogens upon Advanced Malignant Disease', *British Medical Journal*, ii, 4368, 393–8.

Hayward J. 1970, *Hormones and Human Breast Cancer: An Account of 15 Years Study*, London: Heinemann.

Hayward J. 1966, 'Assessment of Response to Treatment at Guy's Hospital Breast Clinic', in J. Hayward and R. Bulbrook (eds) *Clinical Evaluation of Breast Cancer: Proceedings of the First Imperial Cancer Research Fund Symposium, 1965*, London and New York: Academic Press, 131–40.

Hayward J. and Bulbrook R. 1965, 'The Value of Urinary Steroid Estimations in the Prediction of Response to Adrenalectomy or Hypophysectomy', *Cancer Research*, 25, 1129–34.

Heuson J.-C. 1974, 'Hormones by Administration', in H. Atkins (ed.) *The Treatment of Breast Cancer*, Lancaster: Medical and Technical Publishing, 113–64.

Howell A., Clarke R. B. and Anderson E. 1997, 'Oestrogens, Beatson and Endocrine Therapy', *Endocrine-Related Cancer*, 4, 371–80.

Huggins C. and Bergenstal D. 1952, 'Inhibition of Human Mammary and Prostatic Cancers by Adrenalectomy', *Cancer Research*, 12, 134–41.

Irvine W. 1964, 'Selection and Treatment of Patients with Advanced Breast Cancer', *Clinical Radiology*, 15, 161–7.

Irvine W. T., Aitken E. H., Rendleman D. F. and Folca P. J. 1961, 'Urinary Oestrogen Measurements after Oophorectomy and Adrenalectomy for Advanced Breast Cancer', *Lancet*, 278, 7206, 791–6.

Jensen E. V., Block G. E., Smith S., Kyser K. and DeSombre E. R. 1971, 'Estrogen Receptors and Breast Cancer Response to Adrenalectomy', in T. Hall (ed.) *Prediction of Response in Cancer Therapy: Monograph 34*, Bethesda: National Cancer Institute, 55–70.

Joslin C. and Gleave E. (eds) 1970, *The Clinical Management of Advanced Breast Cancer: Second Tenovus Workshop, Cardiff 1970*, Cardiff: Alpha Omega Alpha Publishing.

Juret P. 1966, 'Assessment of Response at the Institute Gustave-Roussy', in J. Hayward and R. Bulbrook (eds) *Clinical Evaluation of Breast Cancer: Proceedings of the First Imperial Cancer Research Fund Symposium, 1965*, London and New York: Academic Press, 185–93.

King R. and Roberts M. 1979, 'The Use of Steroid Receptor Assays in Determining Response to Endocrine Therapy: A Summary of the Clinical Data', in

King R. (ed.) 1979, *Steroid Receptor Assays in Human Breast Tumours: Methodological and Clinical Aspects*, Cardiff: Alpha Omega Publishing, 1–6.

King R. J. B., Barnes D. M., Hawkins R. A., Leake R. E., Maynard P. V., Millis R. M. and Roberts M. M. 1979, 'Measurement of Oestrogen Receptors by Five Institutions on Common Tissue Samples', in R. King (ed.) 1979, *Steroid Receptor Assays in Human Breast Tumours: Methodological and Clinical Aspects*, Cardiff: Alpha Omega Publishing, 7–15.

King R. (ed.) 1979, *Steroid Receptor Assays in Human Breast Tumours: Methodological and Clinical Aspects*, Cardiff: Alpha Omega Publishing.

Lerner B. 2001, *The Breast Cancer Wars: Hope, Fear, and the Pursuit of a Cure in Twentieth-Century America*, Oxford: Oxford University Press.

Lewis M. 2007, *Medicine and the Care of the Dying*, Oxford: Oxford University Press.

Löwy I. 1995, '"Nothing More to be Done": Palliative versus Experimental Therapy in Advanced Cancer', *Science in Context*, 8, 209–29.

Löwy I. 1996, *Between Bench and Bedside: Science, Healing and Interleukin-2 in a Cancer Ward*, Cambridge: Harvard University Press.

Luft R. and Olivecrona H. 1957, 'Hypophysectomy in the Treatment of Malignant Tumours', *Cancer*, 10, 789–94.

Marks H. 1992, 'Cortisone, 1949: A Year in the Political Life of a Drug', *Bulletin of the History of Medicine*, 66, 419–39.

Marks H. 1997, *The Progress of Experiment: Science and Therapeutic Reform in the United States, 1900–1990*, Cambridge: Cambridge University Press.

McGuire W. L., Carbone P. P., Seard M. E. and Esche G. C. 1975, 'Estrogen Receptors in Human Breast Cancer: An Overview', in W. L. McGuire, P. P. Carbone and E. P. Vollmer (eds) *Estrogen Receptors in Human Breast Cancer*, New York: Raven Press, 1–7.

Miller A. B., Hoogstraten B., Staquet M. and Winkler A. 1981, 'Reporting Results of Cancer Treatment', *Cancer*, 47, 207–14.

Paterson P. 1936, 'The Effect of Oophorectomy and Splenectomy on Cancer of the Breast and Uterus', *Lancet*, 227, 5886, 1402–4.

Priestman T. and Baum M. 1976, 'Evaluation of Quality of Life in Patients Receiving Treatment for Advanced Breast Cancer', *Lancet*, 307, 7965, 899–901.

Quirke V. 2005, 'Making *British* Cortisone: Glaxo and the Development of Corticosteroids in Britain in the 1950s–1960s', *Studies in History and Philosophy of Biological and Biomedical Sciences*, 36, 645–74.

Saunders C. and Fallowfield L. 1996, 'Survey of the Attitudes to and Use of Quality of Life Measures by Breast Cancer Specialists in the UK', *The Breast*, 5, 425–6.

Sengoopta C. 2006, *The Most Secret Quintessence of Life: Sex, Glands, and Hormones, 1850–1950*, Chicago: University of Chicago Press.

Segaloff A. 1966, 'Assessment of Response to Treatment by the Co-operative Breast Cancer Group', in J. Hayward and R. Bulbrook (eds) *Clinical Evaluation of Breast Cancer: Proceedings of the First Imperial Cancer Research Fund Symposium, 1965*, London and New York: Academic Press, 125–30.

Slater L. 2000, 'Industry and Academy: The Synthesis of Steroids', *Historical Studies in the Physical and Biological Sciences*, 30, 443–80.

Slevin M. L., Plant H., Lynch D. Drinkwater J. and Gregory W. M. 1988, 'Who Should Measure Quality of Life, the Doctor or the Patient?', *British Journal of Cancer*, 57, 109–12.

Stewart H. J., Forrest A. P. M., Roberts M. M., Chinnock-Jones R. E. A., Jones V. and Campbell H. 1969, 'Early Pituitary Implantation with Yttrium-90 for Advanced Breast Cancer', *Lancet*, 294, 7625, 816–20.

Subcommittee on Breast and Genital Cancer, Council on Drugs, American Medical Association 1960, 'Androgens and Estrogens in the Treatment of Disseminated Mammary Carcinoma: Retrospective Study of Nine Hundred Forty-four Patients', *Journal of the American Medical Association*, 172, 1271–83.

Timmermann C. 2011, '"Just Give Me the Best Quality of Life Questionnaire": The Karnofsky Scale and the Objectivity of the Subjective in Cancer Chemotherapy', paper given at 'The View from Below: On Standards in Clinical Practice and Clinical Research' Workshop (16 September), Charité Institute for the History of Medicine, Berlin Germany.

Valier H. and Timmermann C. 2008, 'Clinical Trials and the Reorganization of Medical Research in Post-Second World War Britain', *Medical History*, 52, 493–510.

Walpole A. and Paterson E. 1949, 'Synthetic Oestrogens in Mammary Cancer', *Lancet*, 254, 6583, 783–6.

Watkins E. 2007, *The Estrogen Elixir: A History of Hormone Replacement Therapy in America*, Baltimore: Johns Hopkins University Press.

Welbourn R. 1993, 'Endocrine Diseases', in W. Bynum and R. Porter (eds) *Companion Encyclopedia of the History of Medicine*, Volume 1, London: Routledge, 484–511.

Welbourn R. 1970, 'Chairman's Introduction: Trials of Ablative Surgery', C. Joslin and E. Gleave (eds) *The Clinical Management of Advanced Breast Cancer: Second Tenovus Workshop, Cardiff 1970*, Cardiff: Alpha Omega Alpha Publishing, 48–9.

Zubrod C. G., Schneiderman M., Frei E., Brindley C., Lennard Gold G., Shnider B., Oviedo R., Gorman J., Jones Jr R., Jonsson U., Colsky J., Chalmers T., Ferguson B., Dederick M., Holland J., Selawry O., Regelson W., Lasagna L. and Owens Jr A. H. 1960, 'Appraisal of Methods for the Study of Chemotherapy of Cancer in Man: Comparative Therapeutic Trial of Nitrogen Mustard and Triethylene Thiophosphoramide', *Journal of Chronic Disease*, 11, 7–33.

# 8
# Cancer Research and Protocol Patients: From Clinical Material to Committee Advisors

*Peter Keating and Alberto Cambrosio*

As discussed by social scientists in relation to AIDS,[1] neuromuscular diseases,[2] and, more generally, internet discussion groups[3] and biomedical research,[4] patient activism has been on the rise in recent years.[5] Cancer is no exception. Presently, in the US, more than 850 cancer advocacy organisations and associations – most founded within the last 15 years – sometimes cooperate and sometimes compete in an extremely partisan (and increasingly controversial)[6] environment. The most important of the cancer patient advocate groups, the National Breast Cancer Coalition, formed in 1991, now comprises over 600 member organisations and over 70,000 members.[7] In 1993, eight different patient groups established the Cancer Leadership Council in an attempt to reach a consensus on health care reform; the Council now federates 33 groups. By the mid-1990s other associations began to put together their own strategies and increase their political visibility. In 1996, prostate cancer activists met to form the National Prostate Cancer Coalition (NPCC), and a coalition of advocacy groups came together under the name of the Intercultural Cancer Council in early 1995, calling for, among other things, greater participation of minority physicians and patients in clinical trials.[8] Similar trends can be seen in Europe, where groups such as Europa Donna[9] (breast cancer) and its counterpart Europa Uomo (prostate cancer), established respectively in 1993 and 2002, came together in 2003 with close to 200 other organisations under the umbrella of the European Cancer Patient Coalition (ECPC).[10]

Most social scientists who have studied these events have focused on the patient advocacy groups themselves, drawing up typologies of the groups, proposing models of their emergence and transformation, scrutinising their inner tensions and their relations to the medical establishment. In doing so, they have mostly taken for granted the notion of 'patients' as individuals who can 'empower' themselves by acting through an advocacy group. In this article we will adopt a different focus: we examine the rise of advocacy groups as an episode in a complex series of events linked to the changing configuration of cancer research; we simultaneously attach a floating ontology to the notion of a 'patient': it refers to a concrete individual, but also to the patient as an abstract or a collective entity.[11]

Group trajectories structure individual pathways in a variety of ways. Over the past 50 years, cancer patients, for example, have evolved as a collective entity along a number of interdependent lines that have delineated new images and practices for persons suffering from or surviving cancer. The most important transformation in this respect has been the emergence of clinical research, with clinical trials at its core, as the dynamic centre of oncology. Presently, even though less than 3 per cent of US cancer patients (and about 14 per cent in the UK, the record-setting country)[12] participate in clinical trials, virtually all patients are diagnosed, treated and advised according to protocols, be they standard or experimental protocols. The epistemic, political and economic status of the cancer patient within protocols and the protocol production process has been a recurring theme for both patients and practitioners, especially when the time comes to choose which road to take – which path to follow or decision to make – in that embedded series of protocols commonly referred to as the therapeutic process. Patient pathways, in other words, are oriented around protocols and participation in protocol formulation and management has consequently become a routine (and highly politicised) activity for research scientists, practicing oncologists and patient activists.

Our insistence here on the centrality of protocol production and management, and of patient participation in these processes is related to a thesis that we have advanced elsewhere,[13] according to which clinical cancer research and its associated system of cancer clinical trials constitute a new style of practice whose heterogeneous components are held together by protocols that are themselves the outcome of continual tinkering and serial adjustments. As the foregoing suggests, we understand protocols in the field of oncology to include much more than the standard comparison of two treatments. Indeed, whereas the analytical and historical starting point of protocol production in oncology has been and remains the clinical treatment trial, cancer protocols are not limited to this highly experimental genre just as the therapeutic process is not exclusively cure-oriented; diagnosis, prognosis and palliative care are also pursued through multiple and overlapping protocols.

Cancer patients have served not only as the 'raw material' and primary purpose of clinical cancer research but also as the subject of a network of evolving rules, norms, restrictions and ethical and epistemological dilemmas. Within this latter problematic, cancer patients have come to be represented and to represent themselves as activists. Formerly concerned primarily with raising money and 'awareness', patient groups have come to reject the notion that they are the mute objects of therapy, charity and research and have consequently demanded and received a place as participants at the clinical research table. In other words, patient activists and advocates understand that clinical trials and the protocols they generate are an obligatory point of passage for cancer patients and that all patient pathways converge sooner or later on a protocol. This combination of protocols and pathways has resulted in a series of patient 'figures' that do not obey any single classificatory principle. We will thus encounter 'last chance patients', 'patients as a (scarce) resource', 'community patients', 'private patients', 'consumer patients', 'minority patients', 'psychosocial patients' and 'activist patients'. All these categories fall, one way or another, under the super-ordinate category of 'protocol patients'.

Before going any further, we offer several caveats intended both to sharpen our focus and to deflate any undue expectations. We are principally concerned with the emergence of the primary categories used to sort out patients. While these categories both unite and divide, from an analytic point of view their primary work is to isolate and specify. In this respect, there are many ways of dividing up patients. One of these involves clinico-epidemiological and pathological divisions. We will not deal with the latter insofar as they construe patients as bearers of a specific disease such as breast or lung cancer. Rather, in what follows we will concentrate on pre-clinical divisions as they relate to protocols and their routine and experimental deployment. We will thus focus on the evolution of the categories used by clinicians, administrators and patients themselves to partition the domain of the ailing, the ill and the marked. These categories are not mutually exclusive, in the sense that the same individual can simultaneously belong to more than one category. Their articulation, however, can be problematic, in the sense that they often relate to different ways of 'engaging'[14] patients and physicians in medical activities. Finally, let us emphasise the obvious: the purpose of this article is not to introduce the 'patient's point of view' in opposition to the clinician's point of view. It is, to repeat, to sketch out the terrain in which shared and disputed pre-clinical perspectives emerge.

## Transforming the 'patient': From peers to peer review

Prior to WWII, cancer therapy rested primarily on two modalities, surgery and radiation therapy. Chemotherapy emerged as a third modality in the 1950s. The degree to which patients had access to these different modalities and in which combination, if any, varied not only in time but also in relation to the type of cancer, country or even hospital in which treatment took place.[15] For our present purpose, the key feature of the postwar period is the emergence and development of chemotherapy and the resulting centre-staging of clinical experimentation as embodied in clinical trials. Adding organisational to technical novelty, chemotherapy protocols were designed and tested by networks of 'trialists' that came to be known under the name of 'cooperative groups'.[16] Clinical trials did not remain confined to chemotherapy: once (quickly) established, the new style of practice embraced radiotherapy and surgery as part of multimodal treatment protocols.[17] The experimental turn brought by chemotherapy and the issues and problems it generated, created ripple effects that touched upon all dimensions of cancer treatment, including a repositioning of patients.

### 'Total care' or the early psycho-social patient

Not a simple issue of cause and effect, the initial repositioning of patients can be best understood by looking at the notion of 'total care' that accompanied the introduction of the first chemotherapy treatments. The clinical cancer protocols of the immediate post-war era were instituted at a time when the therapeutic gains to be expected were relatively small compared to the research aspects, not to speak of the toxic side effects of the treatments.[18] In those early days, clinical researchers invariably recruited patients from the 'advanced stages' or those 'on their last gasp'. Presumably, those with the least to lose had the most to gain. This calculus did not garner universal acceptance. Many clinicians found aggressive therapy repugnant and suggested that it might be better to 'let them die in peace'.[19] Clinical researchers consequently attempted to shed positive light on drastic measures such as chemotherapy and the invasive ancillary investigations by framing the undertaking as a subsidiary component of a larger enterprise known as 'total care'. The latter shifted chemotherapy and other therapeutic strategies out of the realm of human experimentation by combining it with enhanced supportive or palliative care.

The original targets of 'total care' were children suffering from leukaemia, although parents were quickly drawn into the circle of concern.[20]

The originator of the term, Sidney Farber,[21] had developed the notion to accompany his free clinic for childhood leukaemia patients. Total care signified that, in these cases, the cancer clinician treated the family as a whole and its psychosocial and economic needs had to be factored into early protocols. Later, more restrictive uses of the term would limit the notion to the combination of specific and supportive therapy. This early form of the 'psychosocial patient' should not be conflated with the more recent, professionalised version of the psychosocial patient we will discuss in a subsequent section.

Despite the backgrounding of the research component of the therapy, total care often generated controversy amongst clinicians since no clear-cut line could be drawn between specific chemotherapeutic effects and supportive therapy. In the 1960s, following the constitution of successful protocols for the treatment of childhood leukaemia, the notion of total care morphed into the concept of 'total therapy'. Retrospective accounts sometimes blend the two[22] but there was no direct conceptual connection. Donald Pinkel, a student of Sidney Farber, initially assembled total therapy at St Jude's Hospital in Memphis. The 'total' part of the treatment (not care) referred to the fact that a variety of therapeutic modalities such as radiotherapy, chemotherapy and blood transfusion were included in the treatment package.

Notions of 'total care' also shaped the earliest manifestations of lay activism not directed towards fund-raising, an activity dominated in the United States by the American Cancer Society. Indeed, present-day patient advocates often trace their origin back to the 1950s. Although the 'activism' of early activists differs from current activism,[23] the turning point for historians and advocates lies in the activities of a breast surgery patient, Terese Lasser, who, in 1954, began offering her services (occasionally unsolicited) to women recovering from breast cancer surgery. Lasser perceived her endeavour as a form of 'total care' for adult women and the targets of her interventions as 'peers in need of education'. A popular guide to women's history has described her activities thus:

> As soon as Lasser discovered how to cope with all these matters [concerning recovery from breast cancer surgery], a compulsion came over her to teach what she knew. She slipped into the hospital rooms of 'mastectomees', bearing gift boxes with: 1) a starter 'falsie' to pin on the inside of their nightgowns, 2) a ball, string, and instructions, which she demonstrated, for the painful but effective arm-restoring exercises that she herself had devised out of her desperation and

genius, and 3) perhaps most daring, her famous 'Letter to Husbands' about sex. Lasser maintained a pretence that she made these calls only at the request of the patient's doctors or family ..., however, she was often escorted out of the front door of Memorial Hospital when she was found visiting patients at random and without the consent of the responsible surgeon.[24]

Despite initial resistance, Lasser developed her activities into a program (called R2R for Reach to Recovery) that gained such widespread international popularity that, in 1969, the American Cancer Society made it its own.[25]

### From 'peer educators' to cancer activists

Given the educational nature of the program, the original 'peers' were not, in fact, peers in the common-sense notion of the term. Not only did the wealthy socialite have a teaching to impart; Lasser had a position, characterised as that of an 'imperious matriarch' in the aforementioned history text, from which to deliver that teaching.[26] The peer as intrusive matriarch underwent a number of substantial modifications in the years that followed. By the early 1970s, volunteers underwent formal training in order to become 'peers'. Visits were not structured according to lay perceptions; instead, they were ordered according to the type of medical intervention: volunteers, for instance, were trained to make 'lumpectomy, mastectomy, reconstruction, and recurrence visits'.[27] As with all cancer-related activities, self-help groups and other forms of peer support have more recently been explicitly drawn into the realm of protocol production and management insofar as they are now the subject of randomised clinical trials to determine their efficacy.[28]

At the same time, and as protocols have taken the centre stage in oncology, R2R members and 'peer educators' in general now recruit cancer patients for clinical trials by 'working proactively in their communities to share general information about cancer clinical trials in a non-threatening, non-coercive environment, moving people to a more positive receptive posture'.[29] Instituted on a worldwide basis, the R2R programme is not without its critics, particularly since the 1990s when a renewed activism amongst breast cancer patients became sharply critical of the normative nature of 'falsies' and other components of the program. Some breast cancer activists are consequently sceptical of the R2R movement in its present form and some have gone so far as to characterise it as an 'object of ridicule and anger among women with breast cancer'.[30]

While radically different positions such as those taken by 'peers' and 'activists' may seem irreconcilable, they have converged on the platform of protocol production and management. The author of the previous remarks, Sharon Batt, for example, began her breast cancer career as an investigative journalist and, following breast cancer diagnosis, founding member of the activist Breast Cancer Action Montreal (BCAM). Like R2R, BCAM developed out of personal experience bringing together Batt and several other women suffering from breast cancer and developed into an educational and advocacy project. The difference between R2R and BCAM, however, was that members of the latter sought to disseminate the latest research findings and to enter into discussion with breast cancer researchers as opposed to divulging information about the many 'faces' of recovery. This epistemic posture is now widely shared amongst peer groups, activists and advocates and forms a common ground of sorts. At the very least, like some members of the R2R program and other cancer activists since the beginning of the new century, Batt has been sitting on numerous grant review committees for breast cancer research. This, then, is the question: what is the road that led cancer patients from peer group to peer review and led peer counselling to include recruitment to clinical trials?

## Why did patients become so active?

The standard story of this transformation of patient attitudes begins in the field of AIDS activism. Following early positive results with the antiretroviral drug AZT, AIDS activists sought and obtained permission to participate in debates held by the AIDS Clinical Trials Group (ACTG) organised by the National Institute of Allergy and Infectious Diseases (NIAD). By confronting producers and gatekeepers of certified knowledge concerning AIDS both at the US Food and Drug Administration (FDA) and within the ATCG, AIDS activists managed to gain several concessions concerning the conduct of clinical trials and the early distribution of drugs.[31] Ilana Löwy has already questioned the extent to which the 'concessions' actually represent novelties within clinical trial practices and has shown that much of what passed for new was already practiced in the field of clinical cancer trials.[32] But she and most other observers agree that the participation of patients in deliberations concerning protocol formulation and, later, the funding of clinical trials was certainly unprecedented.

Among the first to notice this sea-change in attitude were breast cancer activists who promptly demanded and received greater funding for breast cancer research and a greater voice in the setting of funding priorities.[33] This epistemic activism quickly spread throughout the clinical cancer community so that within a decade after the first AIDS activists began participating in protocol deliberations, practically all anatomic categories of cancer had patient representatives at the various levels of protocol formulation from clinical trial monitoring committees to grant reviews and priority-setting exercises.[34] How this happened is often described in terms of a change of consciousness or 'consciousness raising'. As a form of explanation, such a response is more a restatement of the transformation than an analysis of the events leading up to the change. No one doubts that attitudes have changed. The conditions for change, however, raise larger questions. In what follows, we will look somewhat farther afield in the realm of clinical research practice and politics for an understanding of present-day patient participation in clinical trial activities.

### The patient as a scarce resource

At the end of the 1970s, a new group of professionals, independent from the academic cancer centres[35] and from the trialists' cooperative groups – whose trainees they often were – gained access to the process of protocol production in the field of cancer. This process was closely linked to the emergence of a cancer patient scarcity during that same period. The scarcity emerged in a somewhat roundabout way through a confluence of factors. One of the results of the 1971 National Cancer Act was to broaden training in medical oncology. Consequently, an increasing number of cancer patients came to be treated in community hospitals and were thus, in practice, unavailable for research protocols. At the same time, the rise of the large-scale Phase III clinical trial in the 1970s put considerable demands on what was becoming a relatively scarce resource. Continual improvements in diagnosis and pathology also meant continual subdivision and refinement of clinical categories. Clinical trials proliferated as they tried to explore the different effects of agents targeting, alone or in various combinations, an expanding number of disease sub-entities. As too many trials chased too few patients, patient recruitment began to stall. By the beginning of the 1980s, the patient recruitment problem had become critical and its impact widely felt. The problem of patient accrual – whereby patients began to be regarded (and possibly to conceive of themselves) as a 'resource' (and a scarce one, at that) for clinical researchers rather than as passive consumers of cancer care – became a major item on the trialists' agenda and one of their targets became community-based oncologists who hesitated to put patients on trials claiming that their patients did not want to participate in clinical trials.

In order to attract patients to clinical trials the National Cancer Institute (NCI) developed a number of strategies, including an attempt to broaden geographic coverage and thus increase the patient catchment basin, and an attempt to go to the patients by spreading clinical trials to the community level. Reception of the strategy was mixed. Edward Henderson of the Roswell Park Cancer Center, one of the oldest cancer centres in the United States, suggested that as more medical oncologists gravitated into the community hospitals, 'centers may not have access to patients for research'. Emil Freireich of the M.D. Anderson Center concurred adding that 'it would be tragic if all we get out of clinical research is that A is better than B. Centers [unlike community hospitals] have regular interaction with basic scientists at all levels. It would be particularly cruel if a diminishing budget leads to emphasis on straight clinical research without interaction with scientists'.[36] As Paul Carbone of the University of Wisconsin pointed out, however, it would be unwise to dismiss the community hospitals as irrelevant to research:

> You can't escape the fact that we're training more people to go out into the communities. It would be very bad for this group to go on record that research should be done only in centers and that all patients should go to centers. We have a unique opportunity in centers with a critical mass of multidisciplinary skills, to cull out some patients. We should maintain communication with the community people, bring them in, work with them, and pretty soon you'll start getting the patients you want. If not, they'll go to the NCI on their own, and set up a community clinical cancer council. There are certain key clinical biological questions that can be answered in community hospitals, and some questions that can be answered only in centers.[37]

For oncologists practicing at the community level, the establishment of clinical trials in community hospitals offered a means to stay abreast of developments in cancer therapy and to develop some research experience. Patients and protocols therefore became elements of an exchange and as a result of this exchange the NCI deterritorialised the patients in the sense that patients no longer necessarily belonged to the region in which the 'community' was installed. In principle, of course, geography still mattered. As explained by the Director of the Division of Resources, Centers and Community Activities (DRCCA), Peter Greenwald, 'if you are in Los Angeles and your research base is a cancer center, we would expect it to be one of the two comprehensive centers in Los Angeles'. Yet, as another NCI official (Vincent DeVita) subsequently pointed out, 'if you had a great source of breast cancer patients and there was no need for them in your area, there would be no restriction against affiliating with a national cooperative group'.[38]

By establishing an increasingly intimate link between research and treatment, protocols became omnipresent in the trajectory of cancer patients, drawing both the latter and their community physicians within their purview. This process precipitated a number of confrontations, one, for instance, related to the creation of the Community Clinical Oncology Program (CCOP) by the NCI. Established mainly to improve patient recruitment,[39] the CCOP had originally been slated to fall under the control of the cooperative groups and the cancer centres. In 1982, NCI executives decided to separate 'lead institutions' from 'research bases'. Under the new rules, the groups and the centres would remain necessary points of reference as 'research bases', but they would be unable to compete with community hospitals to become lead institutions in the program.[40] By making association with a research base mandatory for a CCOP institution, NCI officials expected that the result would be 'a steady and sufficient flow of patients into cooperative group trials'.[41] The CCOP, meanwhile, successfully sought further autonomy from the groups and centres by requesting and obtaining control of funds used to support the 'research bases'.[42] There was nothing underhanded about these manoeuvres. The Request for Applications clearly laid out the issues involved in the patient recruitment saga.

> In this country, over 80% of patients with cancer are treated in primary care community hospitals and clinics close to their homes. The remainder are treated in university and government hospitals and cancer centers. Currently, the NCI Division of Cancer Treatment supports a national clinical trials program largely through academic centers. ... The past decade has seen increasing numbers of highly trained clinical cancer specialists, experienced cancer specialists in clinical research and protocol care, enter private practice in the community. ... Experience with several cooperative groups has indicated that cancer physicians in community practice produce clinical research data of similar high quality to that of academic centers. Coupled with this growing community expertise in the ability to perform clinical trials, is a need for increased accrual of patients seen primarily in the community setting into high priority national clinical trials.

... Under this procurement, NCI will seek to meet the needs of community cancer patients nationwide, utilize the trained specialist now practicing in community hospitals and clinics and facilitate its own clinical research goals by establishing a system of 100–200 community clinical oncology programs with national distribution.[43]

While the community clinical oncology program targeted community needs, it also sought to fulfil research objectives. It was, in other words, a *quid pro quo*. The cooperative groups had the technology and the clinical protocols. The community hospitals had the patients. In return for access to the patients, the groups would give community oncologists some say in the development of protocols and the conduct of clinical trials. From the point of view of future or generic patients, this exchange could be justified by reference to the goal of improving treatment: a 'good' treatment could only ever be temporary, and the only way to find a better one was to enrol patients in clinical trials. Yet, once a protocol became 'standard treatment', concrete, individual patients could stand in the way. As Gail Katterhagen, Chairman of the National Cancer Advisory Board Cancer Control and Community Subcommittee explained, 'because there is currently an excellent protocol for Hodgkin's, it is being widely used by community oncologists on many patients they see. However, this minimizes the number of patients available for clinical research and, thus, prevents us from developing new, more effective therapies that will cure a higher percentage of patients'.[44]

Non-research community oncologists and research-oriented oncologists in the nation's cancer centres thus entertained a potential conflict, a sort of proxy conflict that foreshadowed patient activists' criticisms of clinical research in the 1990s. The conflict was this: was the point of clinical cancer research and clinical cancer trials to improve patient care or was it to improve knowledge of cancer? While at some level the two cannot be separated, the issue has been and remains centred on how to set the threshold between the two. At the practical level, as we just saw, one of the related issues was whether or not community hospitals should be allowed to participate in clinical trials even though the community oncologists were not strictly speaking clinical researchers. At another level, the issue became one of who best served or represented the patients' interests.

## Representing the (psycho-social) patient

In 1974, 'a small group of physicians seeking to dispel the myth that community physicians were uninterested in and incapable of participation in state-of-the-art cancer care'[45] created the Association of Community Cancer Centers (ACCC). As the name suggests, these were not research organisations, but treatment centres located within community hospitals and populated, for instance, by members of the newly-formed specialty of medical oncology but also by an increasing array of cancer-related ancillary professions. Practicing oncologists claimed to represent the interests of the 85 per cent of patients treated in the community centres, as opposed to the 15 per cent treated in the academic cancer centres. In the late 1970s, ACCC representatives had begun to portray themselves as the self-appointed guardians of cancer patients. According to one of the members of the ACCC, '[o]ne of [President] Kennedy's aides asked me who the ACCC represents. I said we represent cancer patients in community hospitals, 85% of all cancer patients, and that they're too sick to come here for themselves so we're talking for them'.[46] Rather than taxing clinical research as unresponsive to patient needs, the ACCC argued that it was mostly irrelevant by pointing out that '[t]he development of new treatment interventions that can only be applied in the comprehensive cancer setting fails to address the needs of 85% of all cancer patients'.[47]

Having claimed the status of a patient representative, the ACCC went on to lobby for an amendment to the Cancer Act that would require that at least two members of the National Cancer Advisory Board be practicing physicians who actually treated cancer. The ACCC was partly successful although they were unable to make it mandatory that the physicians also be community oncologists. As a result, university and cancer centre physicians became the only practicing members of the Board. To satisfy ACCC demands for greater say in protocol production, the NCI Division of Cancer Treatment created a committee of community oncologists to suggest what kind of clinical research would be useful to them.[48] These skirmishes and the organisational rearrangements they occasioned foreshadowed similar debates that took place 10–15 years later with a somewhat different set of actors, for by the 1990s the role of patients' spokespersons was being played by patient advocacy groups fuelled by an expanding basis of 'cancer survivors'.[49]

The patient of the 1990s was not necessarily the same as the patient of the 1960s and 1970s: neither biologically – treatments had by then modified the natural history of the disease and thus also its trajectory – nor socially. Concurrent to and exemplifying this transformation was the ACCC official doctrine that the cancer patient was best treated by a multidisciplinary team consisting not only of the usual combination of chemotherapists, surgeons and radiation oncologists, but also of an expanding group of dedicated paramedical practitioners – oncology nurses, oncology social workers, psycho-oncologists – and of hospital administrators such as program, data and practice managers. Nurses and social workers had been at the forefront of the 'psychosocial turn' that began in the mid-1970s. They were soon joined by the members of the new psycho-oncology specialty: the first psycho-oncology textbook was published in 1989 and the two major specialty journals – *The Journal of Psychosocial Oncology* and *Psycho-Oncology* – were established in 1983 and 1992 respectively.[50] Showing a marked penchant for the development of quantitative tools, psycho-oncologists launched investigations into such questions as whether cancer therapy improved the well-being of patients (quality-of-life studies leading to the actuarial notion of 'quality-adjusted life years') and whether an improved psychological outlook (calculated according to, for example, the Mental Adjustment to Cancer Scale) improved therapeutic outcome.[51] Psychosocial measurements are now routinely included in protocols, and as a result, a new dimension has been added to the space through which patients can be represented.

## Consumer patients

The ACCC's acknowledgement of the role of various kinds of administrators points to a further transformation. The ACCC would increasingly come to see 'their' cancer patients as 'clients' for an emerging 'product line' of oncology services. In the late 1980s, the Executive Director of the Association, Lee E. Mortenson, described, for instance, the oncology 'product line' as 'light years ahead of other areas in the development of product line management'.[52]

The idea of the patient as a (critical) consumer of oncology products can in part be traced back to the overall rise of consumer 'awareness' in the 1960s, beginning with Ralph Nader's celebrated assault on Detroit car manufacturers. But given that there were few, if any, oncology products to consume in the 1960s, a more direct line of influence lies in the emergence of the private patient as a result of changes in American health care financing in the early 1980s that opened the door for oncologists to function as private practitioners and to begin treating cancer patients in their offices. Even though these treatments were merely local adaptations of collectively developed protocols originally established by the NCI cooperative oncology groups operating throughout the nation, the notion of the cancer patient as a consumer operating in a private sector purchasing 'cancer goods' developed throughout the 1980s, namely with the emergence in 1985 of the freestanding cancer centre designed to provide diagnosis and treatment of cancer on an outpatient basis. The free-standing clinics were but a small part of the much larger movement of managed care in the 1980s.

Thus, along the pathway from a scarce (and much sought-after) resource of the 1970s to the advocate of the 1990s, the cancer patient underwent a crucial transformation. In the 1980s, s/he became a *consumer* and, in many cases, oncologists became *service-providers*. Today, such a construal is largely unquestioned in the United States. The 'oncology market' is scheduled to become by 2008 the largest pharmaceutical market.[53] *Fortune Magazine*'s most admired company in 2004 was US Oncology and, as a brief description of the size and activities of the firm attests, the phenomenon of the consumption of oncology products is now deeply embedded in the American system of cancer protocol production.

Having absorbed the Houston-based Oncology Resources and the Dallas-based Physicians Reliance Network in 1999, US Oncology had become the largest for-profit provider of cancer treatment in the United States. Composed of a network of over 900 physicians, US Oncology treats about 15 per cent of all newly diagnosed cancer patients thus gaining access to a significant clinical cancer trial resource. Indeed, the company recruits approximately 1600 patients a year to clinical trials, often financed by pharmaceutical companies. Given the magnitude of the enterprise, it should come as no surprise that US Oncology is also the single largest consumer of therapeutic oncology drugs in the United States.[54] More than a consumer, US Oncology also claims to be a producer of medical knowledge through the conduct of clinical trials. Using a single, 'standardized Institutional Review Board approval process', US Oncology boasts that 'through this highly efficient organization, drugs are sped from the research lab into the real world [through clinical trials], reaching patients as quickly as possible'.[55]

As HMOs, freestanding clinics and insurance companies increasingly came to see cancer patients as consumers of a package of goods to be managed with dexterity, if not to say austerity, clinical research and the resultant clinical protocols came to be seen as yet another item in the overall package. By the end of the 1980s, both clinical researchers and practicing oncologists began to observe a rise in the number of anecdotal reports concerning refusal by third-party payers to assume the clinical costs associated with clinical cancer trials. The problem persisted throughout the following decade although the factual basis that would have allowed one to understand the problem remained elusive and became the subject of numerous information-gathering exercises. The following analysis produced towards the end of the

1990s gives some idea of the complexity of the 'third-party payer problem':

Since the late 1980s, partly in response to the insurance industries' [sic] concern about having to pay for high-cost high-technology therapies such as bone marrow transplantation, almost all health plans have adopted plan language that defines experimental treatment as any treatment that requires an IRB [Institutional Review Board] approval, an informed consent, or is given pursuant to an experimental protocol. Furthermore, all the care rendered to the patient under such treatment is included, not just the investigational intervention. Although actual retrospective denials of reimbursement for a patient enrolled in a clinical trial are rare, the potential for the patient to be left liable for expensive therapy due to this common plan exclusion sends a chill to all potential clinical trial enrolees. Surprisingly, there are few, if any, public cases of where a patient was denied access to a clinical trial by a health plan.[56]

The lack of public cases, however, simply meant that evidence for exclusion had to be sought at another level. As the same author goes on to explain:

Usually the patient and the physician will seek prior approval before proceeding with the study, and when they find that the proposed treatment is not a covered benefit, they elect standard therapy because of the need for immediate treatment and the unlikely outcome of reversing the prohibition in the plan language against experimental protocols. The barriers to care found in the survey by Mortensen *et al.* recorded that 37% of the responding oncologists reported that insurer denial of a plan enrolees participation in a clinical trial [sic].[57]

When the issue had first emerged at the end of the 1980s, the NCI's Cancer Therapy Evaluation Program (CTEP), which organised most of the Phase II and III clinical cancer trials run by the cooperative oncology groups in the US, had organised meetings with researchers, pharmaceutical manufacturers and 'lay communities' in order to develop a consensus statement concerning the problem. In other words, even before the emergence of patient demands in the 1990s, the research community had begun to solicit patient groups in an attempt to save clinical research from overzealous accountants.[58] At the very moment, then, when more trials were needed, the patients themselves threatened once again to become a scarce resource.[59] An evaluation by a panel of experts appointed by CTEP of the work carried out by the cooperative groups in 1986 detected problems in patient accrual in 35 to 80 per cent of protocols, depending upon the disease category reviewed.[60] The spectre raised by insurance companies of introducing a distinction between protocol and non-protocol patients would only have made things worse. Clinical researchers and clinicians countered that all patients are protocol patients for insofar as today's treatment protocol is yesterday's research protocol, the separation is purely logical. In other words, all patients were on protocol; the only difference was that some protocols were new and some were old. Moreover, outside the rarefied atmosphere of insurance claims, access to the latest protocols was increasingly becoming a right, not a choice. This, at least, is one way of understanding the rise of 'minority' and 'special' patients.

### The 'special population' patient

Targeted by HMOs and drug companies as consumers in the 1980s and solicited at the community level as 'resources', cancer patients were also called upon to identify themselves as members of ethnic and racial groups starting with the 'black' cancer patient who developed as a separate entity in the mid-1970s.[61] During the 1980s and at the NCI's bequest, Hispanic and Asian cancer patients joined African-Americans first as 'minorities', then as 'special populations' which were allocated a separate Office within the NCI. The 'special populations' designation was invented as an umbrella term in order to draw both 'race' and 'gender' into the same category. Brought together by the same issue – equal access to the resources represented by clinical research and clinical trials – the identity groups had taken different roads to the same destination.

Like the race problem, gender issues had figured prominently in the 1980s. An alleged lack of focus on women's health concerns and the claim that women, like minorities, had been underrepresented in clinical trials, led to the conclusion that women were receiving inferior health care.[62] This perception was, in turn, buttressed by the conduct of several large-scale clinical heart trials in the late 1970s and 1980s that had studied only men. The National Institutes of Health (NIH) responded in 1986 by revising its instructions to applicants for clinical research grants, urging them to include women in their study populations. The following year, the NIH developed an institutional policy regarding the mandatory inclusion of both women and minorities in research as research subjects. In 1990, a bill reauthorising the NIH (and consequently its institutes such as the NCI) would have rendered the policy statutory. Several compromises and controversies later, in 1993, with the passage of the NIH Revitalization Act, minorities and women became statutory components of clinical trial protocols and thus unavoidable clinical entities: the minority patient and the concomitant 'special populations' are now a permanent fixture of US clinical cancer trials and have their own specific place within protocols.

As a consequence, patients can be 'made up' (in Hacking's sense of 'making up people')[63] not only as psycho-social, but also as ethnic beings. The latter is a 'hybrid' category, as it necessarily blends a socio-demographic and a biological dimension, for, in this specific context, the former is relevant only insofar as it can be translated into the latter.

## By way of (temporary) conclusion: Patients as partners and innovators

In contrast to the early 1970s, when the American Cancer Society (ACS) dominated the cancer advocacy scene, by the mid-1980s there had been a significant change in the number and kinds of organisations that sought to effect change with respect to cancer policy and cancer research and practice. In addition to community oncologists, clinical researchers, insurance companies, oncology corporations, drug companies and 'special populations', an increasing number of 'patient groups' organised along disease lines entered the mix. The 'change in consciousness' that occurred in the late 1980s and early 1990s had in fact been prepared in the previous decades as patients and protocols came to be seen as resources subject to equitable distribution and not as 'scientific methods' (protocols) operating on passive 'raw material' (patients). This change in perspective was not restricted to patients but operated up and down the line that produced, funded, organised, and disseminated protocols. As part and parcel of this transformation and as the competition for research dollars and insurance coverage heated up in the 1990s, a somewhat more confrontational series of groups emerged.

It would however be a mistake to portray the relation between cancer patient advocacy groups and health care practitioners and administrators as necessarily or even frequently adversarial. Past is the (not so distant) time when oncologists would wonder about the motives, representative nature and 'secret agenda' of patient advocates.[64] Far from being looked upon with suspicion, the participation of patients' groups in cancer-related activities is now actively sought after by the various actors directly and indirectly involved in cancer protocol production, management and regulation. To this end, professional bodies have established multiple interfaces with patient groups. For instance, beginning in 2002, the European Society for Medical Oncology added a 'patient seminar' to their annual meeting and created a 'Cancer Patient Working Group' designed to lead from 'patient-physician communication' to 'patient-physician partnership'.[65] 'Partnership' and 'active involvement' have indeed become common tropes when professionals discuss patient involvement in research, in Europe as well as in the US.[66] In return, so to speak, patient advocacy group has been actively promoting measures to increase the enrolment of patients in cancer clinical trials.[67]

State agencies have also multiplied patient interfaces. The European Commission, for instance, has applauded the establishment in 2003 of the ECPC and the launch in 2004, at the European Parliament, of a European Patients' Forum as a 'representative body' to which the Commission could turn.[68] The creation in 1995 of EMEA, the European agency in charge of evaluating drugs, has been greeted not simply as a way of promoting innovation in the pharmaceutical sector, but also of simplifying discussions between regulatory authorities and patient representatives from the different European countries.[69] More controversial,[70] of course, are the relations between pharmaceutical companies and patient groups, whereby the latter, by pressuring for the release of new drugs, act as innovators. In 1999, for instance, gastrointestinal stromal tumour patients involved in Phase I trials with Gleevec – soon to become the prototypical example of a new generation of 'targeted' drugs – formed a patient group known as Life Raft Group and devoted in part to sharing information on clinical trial results and side effects. According to its head, the group 'turned the clinical research world upside down. It can take years for patient trial data to filter down into the general medical community and even longer to reach the general public; the Life Raft Group offers that data in real-time'.[71] The patient group in turn became a pressure group. When Gleevec was about to go to market Life Raft and other patients petitioned the company to speed up production time and acted as a recruitment agent for the future product.

It would be naïve, however, to describe patient groups at the beginning of the new century as the victims of manipulation by medical practitioners and pharmaceutical companies. The 2005 and 2006 editions of the 'Masterclass in Cancer Patient Advocacy' organised by the ECPC, for instance, included, in addition to sessions devoted to more obvious topics as discrimination against cancer patients, sessions discussing the issue of access to cancer drugs and trials, and, in particular, the issue of 'patient group – pharmaceutical company relations'.[72] Other skills to be learned by ECPC members and member organisations were effective ways of lobbying national governments and agencies. In other words, cancer patient groups, as in the case with the muscular dystrophy group discussed by Vololona Rabeharisoa,[73] are presently engaged in a form of collective mobilisation that can be best characterised as one in which the group, far from simply representing the interests of sick individuals, mediates between different social actors and institutions. From a historical point of view, this means that if we want to understand the development of cancer patient advocacy we have to consider all the components of the complex network of institution and practices that define the present-day configuration of clinical cancer research.

## Notes

1 Epstein 1996; Barbot 2002; Dodier 2003.
2 Rabeharisoa 2003; Rabeharisoa 2006.
3 Akrich and Méadel 2002; Akrich and Méadel 2007.
4 Rabeharisoa and Callon in Jasanoff (ed.) 2004; Duckenfield and Rangnekar 2004.
5 A PubMed search for articles indexed with the MeSH keyword 'patient advocacy' (introduced in 1976) found an average of 453 articles per year during the first half of the 1980s, 978 per year during the second half of the 1990s and 497 per year in the 2001–2006 period. While possibly an indexing artefact, the marked increase in media attention during the 1990s most likely corresponds to the expansion of patient activism during that decade, followed by a 'backgrounding' of the phenomenon.
6 See Marshall and Aldhous 2006 for the charge that patient groups are becoming extensions of pharmaceutical companies' marketing departments.
7 Visco 1998; Visco 2007; see also Dresser 2001.
8 See Epstein 2007 on the inclusion of 'minorities' in clinical trials.
9 Buchanan *et al.* 2004.
10 Rice 2004a; see also Thornton 2006.
11 See Bourret 2005 on the floating ontology of 'patients'.
12 Sinha 2007, p. 420; for a discussion of low participation rates, see Castel *et al.* 2006.
13 Keating and Cambrosio 2007.
14 See Thévenot 2006 for a discussion of 'regimes of engagement'.
15 For a comparative overview of the development of cancer treatment modalities, see Pickstone 2007.
16 Keating and Cambrosio 2002.
17 It is thus not surprising to find several radiotherapists among the first presidents of a *prima facie* medical oncology cooperative group such as the European Organization for Research and Treatment of Cancer (EORTC). For a discussion

of how trialists turned chemotherapy from a last-resort into a first-line treatment (in conjunction with other modalities) see Cambrosio *et al.* in Gaudillière and Hess (eds) 2007.

18  As a leading oncologist recently noted (Sikora 2007), 'it is so difficult to imagine how the early human experiments could ever have been carried out in today's [2007] ethical climate. Pumping little children full of horrible drugs to obtain just a few weeks survival benefit is no fun for any doctor. But without those pioneers and the suffering children, we would not have the drugs for cancer we routinely use today'.

19  Mercer 1999, p. 409.

20  The emergence of chemotherapy redefined the contours of cancer: in the US, leukaemia – a disease that had been relatively neglected in comparison to, say, breast cancer – became the focus of chemotherapeutic research and a 'model disease' for cancer; see Pickstone 2007, p. 185; Keating and Cambrosio 2002. In the UK, leukaemia had been similarly neglected and had no national charity until 1960 when the parents of a young leukaemia victim, appalled by the prevailing treatment and the lack of medical research into the disease, launched the campaign that led to the foundation of the Leukaemia Research Fund (LRF). Medical professionals quickly took the further development of the LRF into their own hands; see Piller 1994.

21  Farber 1951.

22  E.g., Christie and Tansey (eds) 2003, p. 14.

23  Collyar 2005, see also Timothy 1980 and Lerner 2001.

24  Seaman 1998, p. 70.

25  Boehmer 2000, p. 8.

26  Seaman 1998, p. 69.

27  Willits 1994.

28  Dunn *et al.* 2003.

29  Cauhan and Eppard 2004.

30  Batt 1994.

31  Epstein 1996.

32  Löwy in Lock *et al.* (eds) 2000, pp. 49–81.

33  Dresser 2001.

34  Thornton 2006.

35  On the history of US cancer centres, see Nathan and Benz 2001.

36  Anonymous 1978. Freireich echoed comments made by the co-chair of the Association of American Cancer Institutes' Task 10 Committee, Alvin Mauer, who, when listing the ten 'special attributes of clinical research in cancer centers' noted that 'a lot of data is collected that has nothing to do with whether A is better than B but with patterns, treatment complications, biology of cancer, long term treatment complications' (*Ibid*).

37  Anonymous 1978.

38  Anonymous 1982c.

39  The Program had also found partial justification in the 'diffusion hypothesis', according to which physicians who placed some of their patients on protocol would, when the protocol proved superior, offer the same treatment to their other patients. Upon evaluation following the first round of CCOPs, the hypothesis proved to be false. That, of course, was hardly fatal to the programme as it had only ever been offered as an interesting side-benefit; see Anonymous 1989.

40  Anonymous 1982a.
41  Anonymous 1982b. By 1982, the six cooperative groups that had been funded to extend their clinical trials into community hospitals drew 40 per cent of their patients from the communities.
42  Except in those cases where the group or centre managed to affiliate itself with several community groups and thus make it more efficient to pay the group or centre directly; see Anonymous 1982b.
43  Draft Request for Application, reprinted in Anonymous 1982b. For the official text, see Anonymous 1982e.
44  Anonymous 1982d.
45  'History of ACCC: ACCC's First Thirty Years, 1974–2004', www.accc-cancer. org/ACCC/accc_history.asp (last accessed March 2007).
46  Anonymous 1979.
47  Anonymous 1979.
48  Anonymous 1979.
49  Around 2002, they numbered about eight million in the US; see Holland 2002, p. 216.
50  Holland 2002, pp. 213–14.
51  For an insider's history of psycho-oncology, see Holland 2002.
52  Mortenson in Engstrom *et al.* (eds) 1988, p. 201.
53  Chabner and Roberts 2005, p. 70.
54  Fintor 1999, p. 1273.
55  http://www.usoncology.com/OurServices/USONResearch.asp (last accessed June 2007).
56  Wade 1999, p. 537.
57  *Ibid.*
58  Friedman 1989, p. 613. See also McCabe and Friedman 1989.
59  *Report of the National Cancer Institute Clinical Trials Program Review Group*, August 26, 1997, pp. 14, 24. Other organisations have contributed to the discussion. In 1999, for example, the National Cancer Policy Board, created in 1997 by the Institute of Medicine and the National Research Council, proposed criteria by which Medicare beneficiaries would be covered for clinical trials; see Gelband 1999, pp. 5–6. The penury itself is open to debate. Rettig, for example, claims that, at least for Academic Medical Centres, there is no penury of patients; see, Rettig 2000.
60  NCI Division of Cancer Treatment, 'Clinical Investigations Branch', *Annual Report,* Volume 1, October 1, 1985–September 30, 1986, p. 384.
61  This whole issue is analysed in detail in Epstein 2007.
62  Institute of Medicine 1994.
63  Hacking 1986.
64  Baum 1997.
65  Mellstedt 2006/2007, p. 10.
66  Collyar 2005.
67  For the example of the UK-based CancerBaCUP, see Sinha 2007, p. 420. For the US, see Cauhan and Eppard 2004.
68  Rice 2004a, 2004b.
69  Houÿez 2004.
70  For a recent example, see the debate between Kent 2007 and Mintzes 2007.
71  Quoted in Vasella 2003, p. 131.

72  See the ECPC website: http://www.cancerworld.org/CancerWorld/home.aspx?
    id_stato=1&id_sito=9 (last accessed June 2007).
73  Rabeharisoa 2006.

## Works Cited

Akrich M. and Méadel C. 2002, 'Prendre ses medicaments/prendre la parole:
    Les usages des medicaments par les patients dans les listes de discussion élec-
    tronique', *Sciences Sociales & Santé*, 20(1), 89–116.
Akrich M. and Méadel C. 2007, 'De l'interaction à l'engagement: Les collectives
    électroniques, nouveaux militants dans le champ de la santé', *Hermès*, 47, 145–54.
Anonymous 1978, 'AACI Committee Makes Case for Clinical Research in Centers;
    Cancer Control, Group Funds Eyed', *The Cancer Letter*, 4(49), 2.
Anonymous 1979, 'ACCC Seeks Greater Role in Clinical Research Priorities',
    *The Cancer Letter*, 5(15), 3.
Anonymous 1982a, 'NCI Staff Working Group Leans Toward Starting CCOP with
    Community Hospitals as Lead Funding Agencies', *The Cancer Letter*, 8(1), 1.
Anonymous 1982b, 'Agreement Reached on Most CCOP Issues', *The Cancer
    Letter*, 8(4), 2.
Anonymous 1982c, 'NCAB Votes Down Effort to Delay CCOP RFA', *The Cancer
    Letter*, 8(7), 4.
Anonymous 1982d, 'De Vita Accepts "CHOP Like" Cancer Control Elements in
    CCOP', *The Cancer Letter*, 8(11), 2.
Anonymous 1982e, 'RFA No. 10-NIH-NCI-DRCCA Community Oncology Program',
    *The Cancer Letter*, 8(29), 2–6.
Anonymous 1989, 'CCOP Recompetition Approved', *The Cancer Letter*, 15(5), 2.
Barbot J. 2002, *Les malades en mouvements. La médecine et la science à l'épreuve du
    sida*, Paris: Balland.
Batt S. 1994, *Patient No More: The Politics of Breast Cancer*, Charlottetown, P.E.I.:
    Gynergy Books.
Baum M. 1997, 'Who Truly Represents the Needs of the Consumer Diagnosed with
    Breast Cancer? Who Are These Patients' Advocates? How Are They Informed?
    What, if Any, Are Their Secret Agendas?', *European Journal of Cancer*, 33, 807–8.
Boehmer U. 2000, *The Personal and the Political: Women's Activism in Response
    to the Breast Cancer and AIDS Epidemics*, Albany: State University of New York
    Press.
Bourret P. 2005, 'BRCA Patients and Clinical Collectives: New Configurations of
    Action in Cancer Genetics Practices', *Social Studies of Science*, 35, 41–68.
Buchanan M., O'Connell D. and Mosconi P. 2004, 'EUROPA DONNA, the Euro-
    pean Breast Cancer Coalition: Lobbying at European and Local Levels', *Journal of
    Ambulatory Care Management*, 27, 146–53.
Cambrosio A., Keating P. and Mogoutov A. 2007, 'Protocols, Regimens and Sub-
    stances: The Socio-Technical Space of Anti-Cancer Drugs', in J. P. Gaudillière and
    V. Hess (eds) *Ways of Regulating: Therapeutic Agents Between Laboratories, Plants,
    Consulting Rooms*, Berlin: Max-Planck-Institute for the History of Science (Working
    Papers Series).
Castel P., Négrier S. and Boissel J. P. on behalf of the Plateforme d'Aide à la
    Recherche Clinique en Cancérologie de la région Rhône-Alpes, 2006, 'Why

Don't Cancer Patients Enter Clinical Trials? A Review', *European Journal of Cancer*, 42, 1744–8.

Cauhan C. and Eppard W. 2004, 'Systematic Education and Utilization of Volunteer Patient Advocates for Cancer Clinical Trial Information Dissemination at NCCTG Community Sites', *Journal of Clinical Oncology*, 22, 14S, 1031.

Chabner B. A. and Roberts T. G. 2005, 'Chemotherapy and the War on Cancer', *Nature Reviews Cancer*, 5, 65–72.

Christie D. A. and Tansey E. M. (eds) 2003, *Leukaemia* (vol. 15 of *Wellcome Witnesses to Twentieth Century Medicine*), London: Wellcome Centre for the History of Medicine at UCL.

Collyar D. 2005, 'How Have Patient Advocates in the United States Benefited Cancer Research?', *Nature Reviews Cancer*, 5, 73–8.

Dodier N. 2003, *Leçons politiques de l'épidémie de sida*, Paris: Éditions de l'EHESS.

Dresser R. 2001, *When Science Offers Salvation. Patient Advocacy and Research Ethics*, New York: Oxford University Press.

Duckenfield M. and Rangnekar D. 2004. *The Rise of Patient Groups and Drug Development. Towards a Science of Patient Involvement*, London: University College London.

Dunn J., Steginga S. K., Rosoman N. and Millichap D. 2003, 'A Review of Peer Support in the Context of Cancer', *Journal of Psychosocial Oncology*, 21, 55–67.

Epstein S. 1996, *Impure Science: AIDS, Activism, and the Politics of Knowledge*, Berkeley: University of California Press.

Epstein S. 2007, *Inclusion: The Politics of Difference in Medical Research*, Chicago: The University of Chicago Press.

Farber S. 1951, 'The Effects of Therapy on the Life History and Biology of Leukemia', *Proceedings of the Institute of Medicine of Chicago*, 18(14), 311–25.

Fintor L. 1999, 'For-Profit Treatment Centers: Trailblazing A New Model of Care?', *Journal of the National Cancer Institute*, 91, 1272–4.

Friedman M. 1989, 'Summary Report (Associate Director for Cancer Therapy Evaluation)', in *National Cancer Institute, Division of Cancer Treatment Annual Report* (1 October 1988–30 September 1989), 2, 611–61.

Gelband H. 1999, *A Report on the Sponsors of Cancer Treatment Clinical Trials and Their Approval and Monitoring Systems*, Washington D.C.: Institute of Medicine.

Hacking I. 1986, 'Making Up People', in T. C. Heller, M. Sosna and D. E. Wellbery (eds) *Reconstructing Individualism: Autonomy, Individuality, and the Self in Western Thought*, Stanford: Stanford University Press, 222–36.

Holland J. C. 2002, 'History of Psycho-Oncology: Overcoming Attitudinal and Conceptual Barriers', *Psychosomatic Medicine*, 64, 206–21.

Houÿez F. 2004, 'Active Involvement of Patients in Drug Research, Evaluation, and Commercialization: European Perspective', *Journal of Ambulatory Care Management*, 27, 139–45.

Institute of Medicine 1994, Committee on the Ethical and Legal Issues Relating to the Inclusion of Women in Clinical Studies, *Women and Health Research: Ethical and Legal Issues of Including Women in Clinical Studies*, Washington, D.C.: National Academy Press.

Keating P. and Cambrosio A. 2002, 'From Screening to Clinical Research: The Cure of Leukemia and the Early Development of the Cooperative Oncology Groups, 1955–1966', *Bulletin of the History of Medicine*, 76, 299–334.

Keating P. and Cambrosio A. 2007, 'Cancer Clinical Trials: The Emergence and Development of a New Style of Practice', *Bulletin of the History of Medicine*, 81, 197–223.

Kent A. 2007, 'Should Patient Groups Accept Money from Drug Companies? Yes', *British Medical Journal*, 334, 934.

Lerner B. H. 2001, *The Breast Cancer Wars: Hope, Fear, and the Pursuit of a Cure in Twentieth-Century America*, New York: Oxford University Press.

Löwy I. 2000, 'Trustworthy Knowledge and Desperate Patients: Clinical Tests for New Drugs from Cancer to AIDS', in M. Lock, A. Young and A. Cambrosio (eds) *Living and Working with the New Medical Technologies*, Cambridge: Cambridge University Press, 49–81.

Marshall J. and Aldhous P. 2006, 'Patient Groups Swallowing the Best Advice?', *New Scientist*, 28 October, 19–22.

McCabe M. and Friedman M. A. 1989, 'The Impact of 3rd Party Reimbursement on Cancer Clinical Investigation: A Consensus Statement Coordinated by the National Cancer Institute', *Journal of the National Cancer Institute*, 81, 1585–6.

Mellstedt H. 2006/2007, 'The Dawn of a Golden Age of Medical Oncology', *Hospital Healthcare Europe. Pharmacy & Therapeutics*, 9–10.

Mintzes B. 2007, 'Should Patient Groups Accept Money from Drug Companies? No', *British Medical Journal*, 334, 935.

Mercer R. D. 1999, 'The Team', *Medical and Pediatric Oncology*, 33, 408–9.

Mortenson L. E. 1988, 'The Cancer Program Product Line Emerges', in P. F. Engstrom, P. N. Anderson and L. E. Mortenson (eds) *Advances in Cancer Control: Cancer Control Research and the Emergence of the Oncology Product Line*, New York: Alan R. Liss, 199–202.

Nathan D. and Benz E. J. Jr 2001, 'Comprehensive Cancer Centres and the War on Cancer', *Nature Reviews Cancer*, 1, 240–5.

Pickstone J. V. 2007, 'Contested Cumulations: Configurations of Cancer Treatments Through the Twentieth Century', *Bulletin of the History of Medicine*, 81, 164–96.

Piller G. J. 1994, *Rays of Hope. The Story of the Leukaemia Research Fund*, London: Leukaemia Research Fund.

Rabeharisoa V. 2003, 'The Struggle Against Neuromuscular Diseases in France and the Emergence of the "Partnership Model" of Patient Organization', *Social Science & Medicine*, 57, 2127–36.

Rabeharisoa V. 2006, 'From Representation to Mediation: The Shaping of Collective Mobilization on Muscular Dystrophy in France', *Social Science & Medicine*, 62, 564–76.

Rabeharisoa V. and Callon M. 2004, 'Patients and Scientists in French Muscular Dystrophy Research', in S. Jasanoff (ed.) *States of Knowledge. The Co-Production of Science and Social Order*, London: Routledge, 142–60 and 295–6.

Rettig R. A. 2000, 'Are Patients a Scarce Resource for Academic Clinical Research?', *Health Affairs*, 19(6), 195–205.

Rice M. 2004a, 'Formation of Pan-European Groups Buoys Cause of Patient Advocates', *Journal of the National Cancer Institute*, 96, 1498–9.

Rice M. 2004b, 'European Patients Find a Voice', *European Journal of Cancer*, 40, 1285.

Seaman B. 1998, 'Breast Cancer', in W. Mankiller, G. Mink, M. Navarro, B. Smith and G. Steinem (eds) *Reader's Companion to U.S. Women's History*, Boston: Houghton Mifflin, 67–71.

Sikora K. 2007, 'Cancer Case Histories [Book Review]', *Nature*, 447, 641.

Sinha G. 2007, 'United Kingdom Becomes the Cancer Clinical Trials Recruitment Capital of the World', *Journal of the National Cancer Institute*, 99, 420–2.

Thévenot L. 2006, *L'action au Pluriel. Sociologie des Régimes D'engagement*, Paris: La Découverte.

Thornton H. 2006. 'Patients and Health Professionals Working Together to Improve Clinical Research: Where Are We Going?', *European Journal of Cancer*, 42, 2454–8.

Timothy F. E. 1980, 'The Reach to Recovery Program in America and Europe', *Cancer*, 46, 1059–60.

Vasella D. 2003, *Magic Cancer Bullet: How a Tiny Orange Pill is Rewriting Medical History*, New York: HarperCollins.

Visco F. M. 1998, 'The National Breast Cancer Coalition (NBCC)', *Breast Disease*, 10(5/6), 15–21.

Visco F. M. 2007, 'The National Breast Cancer Coalition: Setting the Standard for Advocate Collaboration in Clinical Trials', in S. P. L. Leong (ed.) *Cancer Clinical Trials: Proactive Strategies* (Vol. 132 of *Cancer Treatment and Research*), New York: Springer, 143–56.

Wade J. L. 1999, 'Effect of Managed Care on Community Oncology Clinical Practice and Research', *Cancer Investigation*, 17, 535–42.

Willits M. J. 1994, 'Role of "Reach to Recovery" in Breast Cancer', *Cancer*, 74, 2172–3.

# 9
# Uncertain Enthusiasm: PSA Screening, Proton Therapy and Prostate Cancer

*Helen Valier*

The twentieth century decline in the incidence and prevalence of infectious disease has long been recognised by historians as coinciding with a renewed biomedical focus on the 'disease management' of the chronically ill.[1] During the 1960s and 1970s the use of 'risk factors' – clinical indicators, genetic markers, lifestyle choices, and the like – began to increase the frequency and intensity of similar disease management interventions in seemingly healthy populations. During the past 40 years the global health care industry has engineered hugely profitable markets from healthy 'patients' largely by appealing to the value of preventative intervention in the battle against the new diseases of civilisation: hypertension, cancer, and diabetes.[2] Ilana Löwy, Robert Aronowitz, and Charles Rosenberg have all recently documented some disturbing trends in disease management directed at the aggressive prevention of *anticipated* undesirable outcomes.[3] New diagnostic tools and larger programmes of more biologically sensitive screening have lead to ever greater 'early detection' of 'pre-cancerous', 'pre-diabetic', and 'pre-hypertensive' patient populations. As Aronowitz points out, the experiences and patient-pathways of these 'pre-patient' patients can become almost indistinguishable from those patients with serious clinical symptoms of disease.[4] While the consequences of this elision between statistical *risk* of disease and actual organic illness can be relatively benign, Aronowitz, Löwy and Rosenberg highlight at least one dire consequence of this trend: the rising number of healthy but 'BRCA positive' women undergoing extremely drastic measures such as prophylactic double mastectomies.

The idea that developers of medical technologies and pharmaceuticals 'look for' likely patient populations to diagnose and treat is far from new.[5] Of greater interest perhaps are the numerous case-studies that, when taken on aggregate, seem to suggest that new disease categories might be *routinely* created from the stuff of abnormal cervical smears, mammograms, and blood tests.[6] That the application of new medical technology *routinely* shifts disease management policies designed for the treatment of advanced disease to earlier and earlier 'stages' of (pre)disease states is a startling and perplexing claim. In the case-study I present in this essay, I agree with earlier authors that disease screening can 'create' a pre-disease state (in this case asymptomatic prostate cancer) and thus invoke a disease management intervention. Similarly, I show that disease 'advocates', be they members of the health care industry, or patients themselves, organise to increase awareness and resources so reifying the new 'disease' category. The prostate story does add another dimension to this analysis, however, in that an upsurge in new *screening* technology co-evolved with a new type of anti-cancer *treatment* technology: proton beam therapy. While other studies highlight how pre-patients are managed through interventions designed for the seriously ill, in the case I discuss here, screening for the pre-patient helped create a *new* patient pathway, one that would have consequences for the treatment and management of the seriously ill as well as the apparently healthy.

Here, we have an example of the pre-patient sustaining the growth of a new treatment modality: a technology that continues to expand even as the plausibility of the notion of the pre-patient *as* the patient has come into dispute. As I will show, such pathways can emerge with a promised market (in this case, a commonly diagnosed cancer) and persist due to market and patient enthusiasm: a persistence that seems somewhat immune to the thrall of evidence-based medicine and cost-efficiency analysis.

While the use of charged particles to treat cancer dates back to spin-offs from the Manhattan Project, the commercial exploitation of this technology did not get underway until relatively recently. The 1990s saw a number of highly expensive proton therapy centres built to treat a variety of rare adult and paediatric tumours, and, in the US in particular, a limited number of more common cancers, such as adenocarcinoma of the prostate. The absence of controlled clinical trial data made the introduction of proton therapy into routine use controversial. Historically speaking, the field of radiation oncology is perhaps uniquely burdened by the vast expense of its technological tools as a contributing factor in limited opportunities for clinical trials. A 2004 editorial for the *International Journal of Radiation Oncology, Biology, Physics* (the official journal of ASTRO, the American Society for Radiation Oncology) (since 2008) noted that, 'Our field is particularly vulnerable because purchases of high tech equipment involves enormous resources and the hospital-based pressure to use these resources to generate more revenues can be quite formidable'.[7] In other words, if a hospital is going to pay for a new technology, it expects that technology to come into routine (reimbursed) use.

Past advances in radiotherapy like the Cobalt-60 machines of the 1960s were similarly taken up and promoted by enthusiasts, most notably in the US by Gilbert Fletcher's team at the MD Anderson Cancer Center.[8] With industry primed to exploit new treatment options in radiotherapy, perhaps the only difference 50 years on from Fletcher is the rate of dissemination of innovation. Rapid dissemination might create more opportunities for evidence-based research, but it also creates impressive momentum to continue to expand treatment populations regardless of the evidence from clinical trials. While clinical trials, and the treatment guidelines that flow from them, might be the basis of much current medical practice, these examples show that untested, promising ideas can blossom in the liberal American marketplace, regardless of our supposed deference to the 'gold standard' of clinical trials.[9] For us as patients, our diagnostic and treatment pathways can be as heavily shaped by market forces, physician preference, and popular opinion as by any peer-reviewed, rigorous, statistical analysis.

In the case of proton therapy for prostate cancer then two trends collide: the voracious pursuit of 'early detection' and the creation of new pre-patients, and the entrepreneurial pursuit of a promising technology in the US medical marketplace implemented with little in the way of rigorous trial-based 'evidence'.[10] If we are to better understand the shaping of patient pathways in the late twentieth century, it is critical that we better understand the synergy created between technologies of disease management from detection through to classification and treatment.

## PSA and the 'prostate cancer epidemic' of the late 1980s and early 1990s

The anatomical position of the prostate, deep to the bony pelvis, and the fact that the gland not only surrounds the proximal urethra but also abuts penile erectile tissue has made prostatic resection a perennially difficult surgery.[11] The radiotherapeutic treatment of the prostate with X-rays is also complicated by this deep anatomical positioning, since reaching the desired target area carries great risk of damage to surface and surrounding tissues due to the intensity of irradiation required. For this reason, apart from some brachytherapy, there was historically little radiation treatment of localised prostatic cancer prior to the introduction of mega-voltage Cobalt-60 machines in the 1960s (and later, linear accelerator machines such as IMRT in the 1980s).[12] Endocrine approaches were also beset with problems; orchidectomy or oestrogen therapy became popular from the 1940s onwards but carried obvious, undesirable, side-effects for the men who received this kind of treatment.[13]

In addition to the difficulties of treatment, clinicians have had few tools at their disposal for the diagnosis of prostate cancer. Digital rectal examination of the prostate has been used since the late nineteenth century to reveal any telltale nodularity and hardness in the posterior part of the gland. As needle biopsy became increasingly used during the interwar years, histological research on clinically abnormal prostates followed. Reliable grading of tissue biopsies came with the work of Donald Gleason and his colleagues in the Veteran's Administration Cooperative Urology Research Group during the early 1960s.[14] Despite advances in equipment and technique, needle biopsy of the prostate then and now is a major intervention, carrying risks of serious side-effects, including bacterial infection and bleeding.[15]

Immunological research in the early 1960s produced evidence of antigens specific to the prostate, but it was not until the late 1970s that researchers at the Roswell Park Cancer Institute in Buffalo, New York, turned the investigation of prostate specific antigen (PSA) specifically towards the diagnosis of cancer.[16] In 1987 a team from Stanford University, lead by Thomas Stamey, published a landmark study of PSA in the serum samples from 699 patients, 378 of whom had prostatic cancer.[17] In this study and another published two years later, the team noted elevated PSA levels in those men with malignancies as compared to the non-cancerous group, and a positive correlation between an increased level of PSA and increased tumour volume.[18] Stamey concluded that serum PSA level might become a useful tool for cancer detection, as well as a means to measure the responsiveness of a tumour to anti-cancer therapy, and to monitor for the recurrence of a cancer following treatment. As the technique passed into clinical use, the reported incidence of prostate cancer increased dramatically. Indeed, National Cancer Institute data show a remarkable 16.5 annual percentage change in incidence for the years 1988 to 1992.[19]

Two decades later Stamey and his colleagues published the results of years of follow-up studies, and reflected on the phenomenon of PSA screening:

> virtually all men with prostate cancer can now be detected. On the surface this would appear to be a great epidemiological accomplishment

except for the disturbing fact that while prostate cancer is a ubiquitous tumor, it has an extraordinarily small death rate.[20]

In contrast to their early findings, Stamey and his colleagues now believed PSA levels to show poor correlation to tumour size. Part of this confusion arose because there was indeed a positive relationship between elevated serum PSA levels and increased *prostate* volume. What the researchers found, however, was that the majority of such increases in prostate size were due to benign changes. The fact that such men when subjected to needle biopsy showed malignant cells in their samples proved only how common it was to find malignant cells in otherwise healthy prostates.[21] In a 2004 interview Stamey described how early confusion over the diagnostic benefits of PSA had contributed to a massive overuse of needle biopsies. While acknowledging that for individual men undergoing treatment for prostate cancer, PSA levels retained a great deal of value as a biological indicator of therapeutic response, he urged a swift shift in attitudes linking high PSA level to the requirement for biopsy:

> Any excuse you use to biopsy the prostate – and we've been using PSA as an excuse – you're very likely to find cancer. So the real need, and that's what I have PhDs and MDs in my laboratory working on all the time, is that we need to get a marker for prostate cancer that is proportional to the amount of cancer in the prostate. Then we might be able to make some intelligent decisions about who should be treated and who shouldn't.[22]

Such concerns over the potential for PSA-driven over-diagnosis and over-treatment of prostate cancer were not new. Early in the 1990s two large-scale, randomised, clinical trials were organised to attempt to distinguish the effect of PSA screening on rates of death from prostate cancer. The European Randomised Screening for Prostate Cancer Trial (ERSPCT) randomly assigned 182,000 men aged between 50 and 74 years to either a group that was offered PSA screening every four years, or to a group that was not offered any such screening. When the ERSPCT team reported their findings in 2007 they noted a relative decrease in mortality of about 20 per cent within the screened group but warned:

> The rate of overdiagnosis of prostate cancer (defined as the diagnosis in men who would not have clinical symptoms during their life-time) has been estimated to be as high as 50% in the screening

group ... Overdiagnosis and overtreatment are probably the most important adverse effects of prostate-cancer screening and are vastly more common than in screening for breast, colorectal, or cervical cancer.[23]

In their analyses of ERSPCT results, French biostatisticians Catherine Hill and Agnes Laplanche noted that these high rates of over-diagnosis, combined with the serious medical side-effects suffered by the 50 per cent or so of men treated, should be an overwhelming argument against the use of PSA testing in screening.[24] Similarly, in 2010 a pair of physicians from Case Western and the Cleveland Clinic pressed the point with regard to their own observations:

At this large tertiary care and community medical center, PSA has performed hardly better than a coin toss in predicting prostate biopsy results, regardless of patient age. The controversy surrounding the management of low grade prostate cancers, further magnifies the need for both scientific and ethical scrutiny of PSA and the courage to abandon it as a screening test.[25]

The US counterpart of the European trial – the Prostate, Lung, Colon, and Ovarian (PLCO) Cancer Screening Trial – ran from 1993 to 2001 and enrolled 76,693 men aged 55 to 74 years. For the purposes of the trial, these men were randomly assigned to either an annual screening group and offered PSA testing for six years, and Digital Rectal Exams (DRE) for four years, or to a control group left in the 'routine care' of their regular physicians ('control group' participants also sometimes received a PSA test and DRE, however, depending on what a participant's physician defined as 'regular care', a feature of protocol design that has been criticised).[26] The results of the trial, published in 2010, showed no difference in prostatic cancer mortality between the screened group and the control group at the seven year follow-up stage. The subcommittee tasked with reporting on the effects of PSA screening on quality of life is expected to report in 2013.

For the time being it seems that the US will remain in the grip of 'massive, unwarranted PSA screening'[27] as the popularity of PSA as an oncological marker retains serious market value in US medicine, and as the mantra of early-detection-as-panacea remains embedded in the mind of the American public. Despite the findings of the US and European trials, the American Urological Association continues to endorse PSA screening (accompanied by a digital rectal examination) for men over 40 years old, while the National Cancer Institute (NCI) guidelines note the limitations of the test, but stress its continuing use as a screening tool. In a 2010 opinion piece for the *New York Times* – 'The Great Prostate Mistake' – Richard Albin, one of the original contributors to the identification of PSA, despaired of the reluctance of doctors and their patients to change their views on screening, noting that the annual US bill for this habit was around $3 billion (much of it paid out of the public purse via Medicare and Veterans benefits). 'I never dreamed', he wrote,

> that my discovery four decades ago would lead to such a profit-driven public health disaster. The medical community must confront reality and stop the inappropriate use of P.S.A. screening. Doing so would save billions of dollars and rescue millions of men from unnecessary, debilitating treatments.[28]

A reversal in screening policy could also run into opposition from prostate cancer patient advocates. Patient backlash against perceived criticism of PSA has already been documented. In 2002 two physicians, Gavin Yamey and Michael Wilkes, reported on their experience of publicly challenging a piece in the *San Francisco Chronicle* on the PSA testing and subsequent prostate surgery of local baseball hero, Dusty Baker:

> We wrote to the *Chronicle* arguing that the newspaper had failed to reflect the massive controversy surrounding prostate cancer screening. The *Chronicle*'s editorial team knew nothing about the controversy, which is no surprise given the dominance of the US media by the pro-screening lobby. The editors invited us to write an opinion piece discussing the reasons why men should not be screened. ... Within hours of our piece being published, prostate cancer charities, support groups, and urologists around the country had circulated a 'Special Alert' by email. This community has huge faith in PSA tests, and it did not care for our opinion. The email, under the header 'ATTENTION MEN!!' urged the community to take action. By the end of the day, our email inboxes were jammed with accusations, abuse, and threats. We were compared to Mengele, and accused of having the future deaths of hundreds of thousands of men on our hands.[29]

A 2007 report of the National Cancer Coalition (now renamed Zero: The Project to End Prostate Cancer), 'The Prostate Cancer Gap: A Crisis in Men's Health',[30] speaks to the sense amongst some patient advocates that prostate cancer has been unfairly marginalised in terms of health policies, political resources, and media visibility, especially in comparison to the money and attention devoted to breast cancer research and treatment. As patient advocates have encouraged more men to be open about their emotional and physical health in recent years, so going to get 'the test' has become a symbol of personal responsibility and masculine solidarity. In this charged context then attacks on PSA can be (and are) construed as attacks on gender equity in health care.

A new technique was poised to take advantage of this new trend in disease awareness amongst health-conscious men. The use of proton beams for the treatment of prostate cancer promised fewer of the debilitating side-effects (notably incontinence and impotence) than conventional external beam radiotherapy, and was aggressively marketed at a patient population being diagnosed at earlier stages, and younger ages.

## Proton therapy: From big science to big business

The history of proton therapy as an anti-cancer modality is in essence the history of American biomedicine. Originating in the 'Big Science' of the late interwar period, proton therapy operated within the emergent constellation of high technology, cross-disciplinary, and market-oriented technoscience that came to dominate the medical marketplace by the turn of the twentieth century.[31] The technological basis of proton therapy – the cyclotron – is a product of extensive military, academic, and industrial alliances in high energy and nuclear physics on the decades after the Second World War.[32] Perhaps encouraged by his Quaker background to find humanitarian applications in nuclear technology, Robert Wilson, head of the Manhattan Project cyclotron division suggested that protons might be turned to cancer treatment.[33] Wilson's idea to provide a high dose beam to a tumour with little external scatter into adjacent healthy tissue was and remains the attraction of high-energy treatment.[34] Proton therapy, conducted between physics experiments at Lawrence Berkeley Laboratory, flourished between the 1950s and early 1990s. The scarcity of cyclotrons meant a slow development of applied medical research elsewhere. In the late 1950s, the Harvard Cyclotron Laboratory joined the Lawrence Berkeley Laboratory and the Gustaf Werner Institute in Uppsala, Sweden, in starting a medical programme. The USSR was active in the field beginning in the late 1960s, and Japan began a programme a decade later. It was during the late 1980s and early 1990s that proton therapy facilities became more widespread around the world. England, France, Germany, South Africa, and South Korea joined Russia and Japan in new ventures. One reason for this trend was that new methods in computer tomography, magnetic resonance imaging, and positron emission tomography, allowed for better visualisation and characterisation of tumour target sites and thus more opportunities for applying proton therapy research.[35]

Expense and a dearth of patients initially curtailed efforts to conduct large-scale clinical trials in proton therapy, but in large part the basic physics of particle beams have provided a compelling rationale for incorporating the technology into clinical practice. Accelerated protons exhibit a 'Bragg effect' whereby the energy released by the particle will rapidly increase in its intensity at the limit of its projected pathway. This means that protons can enter the body and pass through intervening tissue with, theoretically at least, only minor disruption before having maximum destructive impact at the targeted tumour site. For tumours in close proximity to critical tissues, such as tumours of the eye, the technique seemed to offer obvious advantages over conventional treatment modalities.[36] Nonetheless, arguments abound over the true extent of these advantages – advantages that are largely supposed rather than proven through clinical trials.

In 1988 the Proton (in 2001 changed to 'Particle') Therapy Cooperative Group (PTCOG) formed for the purpose of advising the Department of Energy in its funding of a joint venture between the Illinois-based Fermi National Accelerator Laboratory (Fermilab) and the California-based Loma Linda Medical Center.[37] The new facility accepted its first proton therapy patient in 1990, and in doing so became the first hospital-based proton therapy unit in the world. Via its newsletter, *Particles*, PTCOG nurtured a growing community of those 'interested in proton, light ion and heavy charged particle radiotherapy', and for those wishing to 'inform radiation oncologists, neurosurgeons, ophthalmologists and others' of the benefits of more widespread use of the modality.[38] Twice yearly meetings kept a still relatively tight-knit group of a few hundred interested oncologists, physicists, and machine vendors, abreast of national and international meetings and new facility development.[39] In 1990 two Loma Linda physicians and PTCOG members, James Slater and Leonard Arzt, launched the much more public 'National Association for Proton Therapy', a not-for-profit advocacy group intended to lobby Congress, professionals, venture capitalists, and the public about the therapeutic potential of the new treatment.[40] New private ventures first in Houston, Texas and then in Jacksonville, Florida soon followed.[41] As of early 2011 there are seven operational proton therapy facilities in the US, with four more in the planning or development stage.[42] This seems to be a trend set to continue as industrial vendors work to produce smaller, cheaper, turnkey cyclotrons.[43]

Proton therapy moved out of Big Science and into big business in the United States thanks in part to two important stimuli.[44] The first was the 1988 approval by the Food and Drug Administration (FDA) of the Loma Linda and Massachusetts General Hospital proton devices; while the second was the successful application by those same institutions to the American Medical Association (AMA) for the creation of proton therapy procedure codes in 1990.[45] FDA approval opened the procedural pathway for new vendors to develop new technologies, while AMA codes provided the financial incentive to manufacturers and providers alike to enter the marketplace.

The limited clinical trial data of the early 1990s showed promise for the use of proton therapy for paediatric and ocular tumours,[46] but both were rare occurrences with very limited patient populations. Prostate cancer, by contrast, was an extremely common cancer, providing a very broad potential patient population. Within three years of opening in 1990, the Loma Linda proton therapy facility had treated 682 patients, 400 of whom were prostate cancer patients.[47] Fourteen years later, a progress report by Loma Linda's Chairman of Radiation Oncology Jerry Slater noted that while the facility treated some 50 different anatomical sites, treatment of adenocarcinoma of the prostate still accounted for some 65 per cent of all patients treated.[48] Loma Linda was not alone in treating this patient population, by the mid-1990s, and in spite of a notable absence of clinical trial data, proton therapy for prostate cancer was an accepted anti-cancer modality in American medicine, popular with consumers, especially men concerned over possible erectile and incontinence problems related to conventional radiotherapy.[49]

For advocates of proton therapy, an insistence on applying the 'gold standard' to proton therapy can be considered not only misguided but also quite unethical. Michael Goiten, a Harvard Medical School radiologist and self-styled 'protoneer'[50] is a strong critic of those who criticise proton therapy:

> It is ... hard to imagine how any objective person could avoid the conclusion that there is, at the very least, a high probability that protons can provide superior therapy to that possible with x-rays in almost all circumstances. It is primarily for this reason that the

practitioners of proton beam therapy have found it ethically unacceptable to conduct RCTs comparing protons with x-rays. Such a comparison would not meet a central requirement for performing RCTs, namely that there be clinical equipoise between the arms of the trial.[51]

The high costs associated with this form of treatment have further stirred the controversy. A 2009 *New York Times* piece by the economics journalist David Leonhardt summarised the problems:

> Some doctors swear by one treatment, others by another. But no one really knows which is best. Rigorous research has been scant. Above all, no serious study has found that the high-technology treatments do better at keeping men healthy and alive. Most die of something else before prostate cancer becomes a problem. 'No therapy has been shown superior to another', an analysis by the RAND Corporation found. Dr. Michael Rawlins, the chairman of a British medical research institute, told me, 'We're not sure how good any of these treatments are'. When I asked Dr. Daniella Perlroth of Stanford University, who has studied the data, what she would recommend to a family member, she paused. Then she said, 'Watchful waiting'.
>
> But if the treatments have roughly similar benefits, they have very different prices. Watchful waiting costs just a few thousand dollars, in follow-up doctor visits and tests. Surgery to remove the prostate gland costs about $23,000. A targeted form of radiation, known as I.M.R.T., runs $50,000. Proton radiation therapy often exceeds $100,000.[52]

The current over-treatment of prostate cancer has helped to fund the growth of proton therapy facilities. The patient population from paediatric and ocular cancers would not fill current capacity,[53] and without large patient populations investors face slim returns. It remains to be seen what effect the current economic crisis will have on the contributions of Medicare to the *status quo*. Perhaps a requirement that proton therapy providers accept reimbursement rates at the levels of comparable therapies will precipitate change.

## Conclusions

Scholarly histories of post-Second World War biomedicine in the US, perhaps particularly in the field of oncology, are generally enthusiastic about the rise of the randomised clinical trial as the 'gold standard' of clinical research. The successful use of clinical trials to investigate treatments against tuberculosis in the 1940s and 1950s and anti-cancer drugs in the 1950s and 1960s spawned a legacy of massive federal investment in large-scale, cooperative, multi-centre, trials in the US, and a sharp growth in the research and development funding of clinical trials by industry. While trials funded from the public purse have often compared treatments already in use, or investigated new combinations of existing treatment modalities, industrial sponsorship of clinical trials has tended to focus more on bringing new drugs to market via approval from the FDA. Specific trials are good fodder for historians, showing as they do moments of important change, consensus emerging from controversy. By contrast, there is relatively little historical attention focused on the decades of critical commentary concerning clinical trials, criticisms that litter the editorial, review, and letters pages of most major medical journals. Historical coverage of the successes and failures of the clinical trial is at best asymmetric and this is important. A focus on success implies that over time medical practice has become increasingly evidence-based. As Iain Chalmers and others have argued, the systematic implementation of fair trials in medicine has been, and remains, a depressingly elusive goal.[54] At a broad stroke, we might worry with good reason that industry-funded research is more likely to show bias towards the publication of positive results,[55] while the government bureaucracy surrounding federal grants leads researchers into years of frustrating red-tape, with trial data frequently abandoned or if the data are published, published against a moving background of ongoing bioscientific research.[56] The gold standard is in crisis.

Meanwhile, as some experts critique the clinical trial and highlight its failings, novel medical interventions can and do enter the market place, dramatically altering patient pathways in the process, with scant regard for standards of evidence. Such findings concern us all if we believe and expect that when we become a patient our own clinical pathway will be rigorously defined, likely benefiting us more than harming us. As taxpayers and consumers of health insurance policies, we might also expect that such pathways provide transparency and cost effectiveness. Commercial promotion, professional enthusiasm, and consumer demand, are all potent indicators of marketplace success. Certainly, our acceptance, as medical consumers, of the 'risky' pre-patient *as* a patient, and our seemingly unshakable belief in the inherent value of screening and early detection create considerable opportunities for aggressive intervention into our ostensibly healthy bodies. The creation of PSA-driven patient pathways in prostate cancer has contributed to an epidemic of over-diagnosis and over-treatment of prostate malignancy in the US, while the increasing use of proton therapy (in preference to conventional forms of radiotherapy) in this patient population has led to dramatic increases in treatment costs. In this case-study, I have argued that the treatment 'shift' between patient and pre-patient is not a one-way phenomenon, but rather a dynamic interrelation. A tool to detect the pre-patient becomes part of the push toward an expensive new therapeutic modality, a modality that will likely have an effect on patient pathways for decades to come.

## Notes

1 Stanton 2000.
2 For an emerging post-Second World War perspective on the increased medicalisation of the healthy, see Lewis 1977; for a more recent view see Greene 2007.
3 Aronowitz 2009; Löwy 2009; Rosenberg 2009.
4 Aronowitz 2009, p. 419.
5 Vos 1991.
6 On the role of diagnostic technology in the creation of new disease categories, see Casper and Clarke 1998; Lerner 2001; Bryder 2008; Aronowitz 2007; and Löwy 2012.
7 Goffman 2004, p. 1.
8 Dahl 2005, p. 798.
9 On the development of the randomised clinical trial as the gold standard of clinical research see for instance, Keating and Cambrosio 2002; Löwy 1996, Chapter 1; Marks 1997.
10 The problem of what counts as 'evidence' in evidence-based medicine is hotly debated; for excellent overviews see the special Spring 2009 issue of *Perspectives in Biology and Medicine*; and several papers in the 2010 no. 3 issue of the *Journal of Evaluation in Clinical Practice* on the nature of evidence and the philosophy of medicine. See also Timmermans and Berg 2003, Introduction.
11 Sriprasad *et al.* 2009.
12 Sriprasad *et al.* 2009, p. 189.
13 Jordan 2009. In recent years, prostate cancer has also been treated with ultrasound and chemotherapeutic agents, but the main treatment for localised tumours remains surgical and radiological, while for metastatic disease, hormone therapy and palliative radiotherapy are the norm.
14 Phillips and Sinha 2009.
15 Silletti *et al.* 2007.
16 Rao *et al.* 2007. Rao notes that the discovery of non-prostatic sources of PSA lead to moves to rename PSA prostate *secreting* antigen in the late 1990s.
17 Stamey *et al.* 1987.
18 Stamey *et al.* 1989, p. 1076.
19 National Cancer Institute 2011. See also, Makarov and Carter 2006, p. 2384; Potosky *et al.* 1995, p. 550.

20  Stamey *et al.* 2004, p. 1300.
21  A 1996 study demonstrated that of 525 men who had died of accidental causes on the streets of Detroit, prostate malignancy typically began amongst men in their 20s, and increased until 80 per cent of men had prostate cancer in their 70s: Sakr *et al.* 1996, p. 138.
22  Barclay 2004.
23  Schröder *et al.* 2009, p. 1328.
24  Hill and Laplanche 2010.
25  Potts and Walker 2010, p. 320.
26  Andriole *et al.* 2009.
27  Stamey *et al.* 2004, p. 1300.
28  Ablin 2010, p. A27.
29  Yamey and Wilkes 2002, p. 431.
30  National Prostate Cancer Coalition 2007.
31  Clarke *et al.* 2003.
32  Halperin 2006, p. 676; Lawrence 1937; Stone and Larkin 1952.
33  Wilson 2004.
34  Hall 2009, p.196. See Tobias *et al.* 1956, p. 121.
35  Metz 2006. In its infancy, proton beams were first directed at the pituitary gland, which could be easily located using X-rays.
36  See for example Suit 2002; Jones and Rosenberg 2005; and Jones and Burnet 2005.
37  *Particles*, no. 2, June 1988. See also Smith 2006, p. R493.
38  *Particles*, no. 1, February 1988.
39  *Particles*, no. 10, July 1992 listed circulation figures at around 100 for its first issue rising to around 470 for the tenth issue five years later. *Particles*, no. 11, January 1993, noted that around 60 per cent of newsletter recipients were in the United States.
40  *Particles*, No. 8, June 1991.
41  Smith 2009, p. 560.
42  The National Association for Proton Therapy 2011.
43  Jones 2005, p. 99. For vendor efforts to produce turnkey technology see for instance the website of a major producer of American proton therapy machines, Advanced Particle Therapy <http://www.advancedparticletherapy.com/> accessed 2/21/2011.
44  Not all proton therapy facilities around the world are capable of treating the same sorts of tumours. The cyclotron at England's Clatterbridge Hospital, for example, is a relatively low energy machine (62 MeV) and provides therapy for intraocular cancers. For more deeply situated tumours in the chest, pelvis, etc., much more powerful cyclotrons are required, with energies over 200 MeV: see Jones 2006. As of 2011, PTCOG lists 37 proton (or heavy-beam carbon) therapy facilities worldwide, 30 of which are high-energy (>200 MeV for protons and >400 MeV for carbon ion) machines: see Particle Therapy Cooperative Group 2011. The US proton therapy movement of the 1990s and 2000s does stand out, however, both for the rapid nature of its growth, fuelled in large part by private investment.
45  Smith 2009, p. 559.
46  See Olsen *et al.* 2007.
47  *Particles*, No. 13, January 1994, p. 6.

48   Slater 2006, p. 85.
49   On the absence of trial data for proton therapy see Turesson *et al.* 2003.
50   Goitein and Goitein 2005, p. 793.
51   Goitein and Cox 2008, p. 175.
52   Leonhardt 2009, p. A1.
53   Kangan and Schultz 2010, p. 409.
54   Evans *et al.* 2006.
55   See for instance, Lexchin 2011.
56   On the perceived 'crisis' in the current publically-funded clinical trials network
     see: Nass *et al.* (eds) 2010, p. ix.

## Works Cited

Ablin R. 2010, 'The Great Prostate Mistake', *New York Times*, 9 March, A27.
Advanced Particle Therapy 2011, 'Partnering for the Future of Cancer Treatment'
   <http://www.advancedparticletherapy.com/> (last accessed 21 February 2011).
Andriole G. L., Grubb R. L., Buys S. S., Chia D., Church T. R., Fouad M. N.,
   Gelmann E. P., Kvale P. A., Reding D. J., Weissfeld J. L., Yokochi L. A., Craw-
   ford E. D., O'Brien B., Clapp J. D., Rathmell J. M., Riley T. L., Hayes R. B.,
   Kramer B. S., Izmirlian G., Miller A. B., Pinsky P. F., Prorok P. C., Gohagan J. K.
   and Berg C. D. 2009, 'Mortality Results from a Randomized Prostate-Cancer
   Screening Trial', *New England Journal of Medicine*, 360, 1310–19.
Aronowitz R. 2007, *Unnatural History: Breast Cancer and American Society*, Cambridge:
   Cambridge University Press.
Aronowitz R. 2009, 'The Converged Experience of Risk and Disease', *Milbank
   Quarterly*, 87, 417–42.
Barclay L. 2004, 'End of an Era for PSA Screening: A Newsmaker Interview With
   Thomas Stamey, MD', *Medscape Medical News:* http://www.medscape.com/view-
   article/489474 (last accessed 20 February 2011).
Bryder L. 2008, 'Debates about Cervical Screening: An Historical Overview',
   *Journal of Epidemiology and Community Health*, 62, 284–7.
Casper M. and Clarke A. 1998, 'Making the Pap Smear into the "Right Tool" for
   the Job: Cervical Cancer Screening in the USA, circa 1940–95', *Social Studies of
   Science*, 28, 255–90.
Clarke A. E., Mamo L., Fishman J. R., Shim J. K. and Fosket J. R. 2003, 'Bio-
   medicalization: Technoscientific Transformations of Health, Illness and US
   Biomedicine', *American Sociological Review*, 68, 161–94.
Dahl O. 2005, 'Protons. A Step Forward or Perhaps Only More Expensive Radiation
   Therapy?', *Acta Oncologica*, 44, 798–800.
Evans I., Thornton H. and Chalmers I. 2006, *Testing Treatments: Better Research
   for Better Healthcare*, London: British Library.
Goffman T. 2004, 'The Vulnerability of Radiation Oncology within the Medical
   Industrial Complex', *International Journal of Radiation Oncology Biology Physics*,
   59, 1–3.
Goitein M. and Cox J. 2008, 'Should Randomized Clinical Trials Be Required for
   Proton Radiotherapy?', *Journal of Clinical Oncology*, 26, 175.
Goitein M. and Goitein G. 2005, 'Swedish Protons', *Acta Oncologica*, 44, 793–7.

Greene J. 2007, *Prescribing by Numbers: Drugs and the Definition of Disease*, Baltimore: Johns Hopkins University Press.

Hall E. 2009, 'Protons for Radiotherapy: A 1946 Proposal', *Lancet Oncology*, 10, 196.

Halperin E. 2006, 'Particle Therapy and Treatment of Cancer', *Lancet Oncology*, 7, 676–85.

Hill C. and Laplanche A. 2010, 'Cancer de la Prostate: Les Données Sont en Défaveur du Dépistage', *La Presse Médicale*, 39, 859–64.

Jones B. 2005, 'Particle Therapy Co-operative Oncology Group (PTCOG 40) Meeting, Institute Curie 2004', *British Journal of Radiology*, 78, 99–102.

Jones B. 2006, 'The Case for Particle Therapy', *British Journal of Radiology*, 79, 24–31.

Jones B. and Burnet N. 2005, 'Radiotherapy for the Future', *British Medical Journal*, 330, 979–80.

Jones B. and Rosenberg I. 2005, 'Particle Therapy Cooperative Oncology Group (PTCOG40)', *British Journal of Radiology*, 78, 99–102.

Jordan V. 2009, 'A Century of Deciphering the Control Mechanisms of Sex Steroid Action in Breast and Prostate Cancer: The Origins of Targeted Therapy and Chemoprevention', *Cancer Research*, 69, 1243–54.

Kangan A. and Schultz R. 2010, 'Proton-Beam Therapy for Prostate Cancer', *Cancer Journal*, 16, 405–9.

Keating P. and Cambrosio A. 2002, 'From Screening to Clinical Research: The Cure of Leukemia and the Early Development of the Cooperative Oncology Groups, 1955–1966', *Bulletin of the History of Medicine*, 76, 299–334.

Lawrence E. 1937, 'The Biological Action of Neutron Rays', *Radiology*, 29, 313–22.

Leonhardt D. 2009, 'In Health Reform, a Cancer Offers an Acid Test', *New York Times*, 7 July, A1.

Lerner B. 2001, *The Breast Cancer Wars: Hope, Fear, and the Pursuit of a Cure in Twentieth-Century America*, Oxford: Oxford University Press.

Lewis T. 1977, 'On the Science and Technology of Medicine', *Daedalus*, 106, 35–46.

Lexchin J. 2011, 'Those Who Have the Gold Make the Evidence: How the Pharmaceutical Industry Biases Outcomes of Clinical Trials of Medications', *Science and Engineering Ethics*, 17, Epub ahead of print.

Löwy I. 1996, *Between Bench and Bedside: Science, Healing, and Interleukin-2 in a Cancer Ward*, Cambridge: Harvard University Press.

Löwy I. 2009, *Preventative Strikes: Women, Precancer, and Prophylactic Surgery*, Baltimore: Johns Hopkins University Press.

Löwy I. 2012, *A Woman's Disease: The History of Cervical Cancer*, Oxford: Oxford University Press.

Makarov D. and Carter H. 2006, 'The Discovery of Prostate Specific Antigen as a Biomarker for the Early Detection of Adenocarcinoma of the Prostate', *Journal of Urology*, 176, 2383–5.

Marks H. 1997, *The Progress of Experiment: Science and Therapeutic Reform in the United States, 1900–1990*, New York: Cambridge University Press.

Metz J. 2006, 'History of Proton Therapy', August 2006, Oncolink.org: http://www.oncolink.org/custom_tags/print_article.cfm?Page=2&id=209&Section=Treatment_Options (last accessed 7 August 2011).

Nass S., Moses H. and Mendelsohn J. (eds) 2010, *A National Cancer Clinical Trials System for the 21st Century: Reinvigorating the NCI Cooperative Group Program*, Washington: National Academy of Sciences: http://www.nap.edu/catalog/12879. html (last accessed 20 February 2011).

The National Association for Proton Therapy 2011, 'US Proton Therapy Centers', http://www.proton-therapy.org/map.htm (last accessed 20 February 2011).

National Cancer Institute 2011, 'Surveillance Epidemiology and End Results. Stat Fact Sheet: Prostate', http://seer.cancer.gov/statfacts/html/prost.html (last accessed 8 August 2011).

National Prostate Cancer Coalition 2007, 'Do it for Dad', http://zerocancer.org/ assets/2007_DIFD_Press_Kit.pdf?docID=661 (last accessed 6 July 2011).

Olsen D. R., Bruland O. S., Frykholm G. and Norderhaug I. N. 2007, 'Proton Therapy – A Systematic Review of Clinical Effectiveness', *Radiotherapy and Oncology*, 83, 123–32.

Particle Therapy Cooperative Group 2011, 'Particle Therapy Facilities on Operation (International)', http://ptcog.web.psi.ch/ptcentres.html (last accessed 21 February 2011).

*Particles*, June 1988–January 1994.

Phillips J. and Sinha A. 2009, 'Patterns, Art, and Context: Donald Floyd Gleason and the Development of the Gleason Grading System', *Urology*, 74, 497–503.

Potosky A. L., Miller B. A., Albertsen P. C. and Kramer B. S. 1995, 'The Role of Increasing Detection in the Rising Incidence of Prostate Cancer', *Journal of the American Medical Association*, 273, 548–52.

Potts J. and Walker E. 2010, 'Isn't It Time to Abandon Prostate Specific Antigen (PSA) for Prostate Cancer Screening?', *Journal of Men's Health*, 7, 3.

Rao A. R., Motiwala H. G. and Karim O. M. A. 2007, 'The Discovery of Prostate-Specific Antigen', *BJU International*, 101, 5–10.

Rosenberg C. 2009, 'Managed Fear', *The Lancet*, 373, 802–3.

Sakr W. A., Grignon D. J., Haas G. P., Heilbrun L. K., Pontes J. E. and Crisman J. D. 1996, 'Age and Racial Distribution of Prostatic Intraepithelial Neoplasia', *European Urology*, 30, 138–44.

Schröder F. H., Hugosson J., Roobol M. J., Tammela T. L. J., Ciatto S., Nelen V., Kwiatkowski M., Lujan M., Lilja H., Zappa M., Denis L. J., Recker F., Berenguer A., Määttänen L., Bangma C. H., Aus G., Villers A., Rebillard X., van der Kwast T., Blijenberg B. G., Moss S. M., de Koning H. J. and Auvinen A. 2009, 'Screening and Prostate-Cancer Mortality in a Randomized European Study', *The New England Journal of Medicine*, 390, 1320–8.

Silletti J. P., Gordon G. J., Bueno R., Jaklitsch M. and Loughlin K. R. 2007, 'Prostate Biopsy: Past, Present, and Future', *Urology*, 69, 413–16.

Slater J. 2006, 'Clinical Applications of Proton Radiation Treatment at Loma Linda University: Review of a Fifteen Year Experience', *Technology in Cancer Research and Treatment*, 5, 81–9.

Smith A. 2006, 'Proton Therapy', *Physics in Medicine and Biology*, 51, R491–R504.

Smith A. 2009, 'Vision 20/20: Proton Therapy', *Medical Physics*, 36, 556–68.

Sriprasad S., Feneley M. and Thompson P. 2009, 'History of Prostate Cancer Treatment', *Surgical Oncology*, 18, 185–91.

Stamey T. A., Caldwell M., McNeal J. E., Nolley R., Hemenez M. and Downs J. 2004, 'The Prostate Specific Antigen Era in the United States is Over for Prostate Cancer: What Happened in the Last Twenty Years?', *Journal of Urology*, 172, 1297–301.

Stamey T. A., Kabilin J. N., McNeal J. E., Johnstone I. M., Freiha F., Redwine E. A. and Yang N. 1989, 'Prostate Specific Antigen in the Diagnosis and Treatment of Adenocarcinoma of the Prostate. II. Radical Prostatectomy Treated Patients', *Journal of Urology*, 141, 1076–83.

Stamey T. A., Yang N., Hay A. R., McNeal J. E., Freiha F. and Redwine E. A. 1987, 'Prostate-Specific Antigen as a Serum Marker for Adenocarcinoma of the Prostate', *New England Journal of Medicine*, 317, 909–16.

Stanton J. 2000, 'Supported Lives', in R. Cooter and J. Pickstone (eds) *Medicine in the Twentieth Century*, Amsterdam: Harwood Academic, 601–15.

Stone R. and Larkin J. 1952, 'The Treatment of Cancer with Fast Neutrons', *Radiology*, 39, 608–20.

Suit H. 2002, 'Coming Technical Advances in Radiation Oncology', *International Journal of Oncology Biology and Physics*, 53, 798–809.

Timmermans S. and Berg M. 2003, *The Gold Standard: The Challenge of Evidence-Based Medicine and Standardization in Health Care*, Philadelphia: Temple University Press.

Tobias C. A., Lawrence J. A., Born J. L., McCombs R. K., Roberts J. E., Anger H. O., Low-Beer B. V. A. and Huggins C. B. 1956, 'Pituitary Irradiation with High-Energy Proton Beams: A Preliminary Report', *Cancer Research*, 18, 121.

Turesson I., Johansson K.-A. and Mattsson M. 2003, 'The Potential of Proton and Light Ion Beams in Radiotherapy', *Acta Oncologica*, 42, 107–14.

Vos R. 1991, *Drugs Looking for Diseases: Innovative Drug Research and the Development of the Beta Blockers and the Calcium Antagonists*, Amsterdam: Kluwer Academic Publishers.

Wilson R. 2004, *A Brief History of the Harvard University Cyclotrons*, Cambridge: Harvard University Press.

Yamey G. and Wilkes M. 2002, 'The PSA Storm. Questioning Cancer Screening Can Be a Risky Business in America', *British Medical Journal*, 324, 431.

# 10

## Patients and their Problems: Situated Alliances of Patient-Centred Care and Pathway Development

*Teun Zuiderent-Jerak, Roland Bal and Marc Berg*

It is 11:00 a.m. on a May morning in a hospital in a large city in the Netherlands. This hospital, which we will call Hospital E, is a university medical center and we are observing the work at the haematology/oncology outpatient clinic and treatment center. Most of the patients here are tertiary referrals who can no longer get the care they need in regional hospitals. Patients are referred by hospitals in the region but also by clinics throughout the Netherlands: particular forms of care offered are known to be of very high quality in Hospital E. We are witnessing a common phenomenon at this time in the morning in the treatment center of the haematology/oncology ward: all the chairs are filled by people receiving chemotherapy and there are still many patients waiting for their consultation with their haematologist or oncologist.

At the counter where new appointments are made there is a patient from the south of Holland who has to travel about 200 kilometers to get the specialized haematological treatment she needs. Together with her husband, she just came from a visit with the junior doctor who gave her an order-sheet indicating that she requires an appointment for a scan at the radiology department, and also a test of her bone marrow. With this sheet they have returned to the counter and have awaited their turn. Fadila,[1] one of the doctors' assistants, is scheduling the requested diagnostics and tries to call a secretary of the radiology department. The secretaries at radiology seem to be very busy and Fadila is put on hold. After holding the line for half an hour, she manages to get through. In this time she couldn't really continue her work at the counter, so the patient records have piled up. This means patients have finished the consult with their doctor, have come back to the counter for following appointments and are waiting for their appointments in the queue or in a seat near the counter.

'Of course you can choose to send patients home, and call them back later to make appointments, but then you shift your work to the afternoon after 2 p.m.', Jane, a colleague of Fadila informs me. I realize that they do not consider this to be a real option since at that time there will be new consults with other patients queuing up.

11:30 a.m.: The patient from the south of Holland hears that her appointment is in a few weeks time, at 8:30 a.m. The husband of this patient (who's doing virtually all the talking since his wife is very tired) indicates that this is quite early if you have to make a three hour journey, and besides, he's not happy that the bone marrow is planned a day before, also at 8:30 a.m. 'But a skeleton is always at 8:30. That can't be arranged differently', Fadila says, 'and I've been working on that appointment for half an hour, and the day before they don't have a place for you in the radiology department'. The patient and her husband leave being rather taken aback, but soon after they return to the counter. The queue has become quite long, and Fadila is helping the next lady. Other patients, two elderly people come to the counter to indicate that they are still sitting there and would like to go home. They are completely worn out. The husband of the patient argues: 'But if there's no place for the scan on the day before, can't we change the bone marrow then?'. Fadila changes this unwillingly.

When the man has left, Fadila says: 'That's so annoying, you know, that I'm trying to get rid of this pile of patient records, and then you have a patient like this constantly wanting something else!'

After some time, again the patient and her husband come back. They have called with their own internist in the south of Holland, who has indicated that they don't have to come to the university hospital for this particular scan. The husband indicates that they have now made an appointment in their regional hospital for this. Fadila is helping another patient, and her colleague takes over the patient from the south of Holland. But Fadila has to interfere anyway, to explain what has just taken place. Now the patient who was being helped by Fadila gets angry with the patient from the south of Holland: 'Well, I'm standing here waiting, and you just jump the queue!'. The husband of this angry woman tries to calm her down.

12:08 p.m.: Fadila is told by the junior doctor in passing that after all the patient needs to get cytogenesis as well, and that she has to make an appointment for this. She does this on the day following the existing appointment for the bone marrow, and informs the patient, who is still waiting near the counter. The husband asks to change the appointment to the same day as the bone marrow. Fadila does this, and also cancels the appointment with the radiology department. This doesn't take half an hour this time, but still takes a long time.

After the patients have left, and the appointments are all made, the doctor comes to the counter. He indicates, 'Well, actually I'm not there when you've made the appointment for the bone marrow, because I'm taking a course then'.

## Introduction: Patient-centred pathways

This situation is unfortunately a rather common sight in Western European outpatient clinics. When presenting this observation to students following a course in health care change management at the institute where we work, a health care professional among them ironically noted that this way of organising treatment is: 'probably the only thing that is properly standardised in Dutch hospital care'.

The observation comes from our research on the organisation of oncology care that we have been pursuing for several years[2] and in this particular research setting, the clinicians and management of the outpatient clinic requested that we analyse the situation of their outpatient clinic, suggest ways to improve the organisation of care and carry out these suggestions together with care professionals and management of the ward. The project set up for this was conceptually related to a hospital-wide experiment with creating standardised care trajectories.

There is much to say about this type of research. We could, for example, discuss the approach we are developing for such studies, which is a blend of ethnographic study of care work and interventionist research practices,[3] or the results that were obtained,[4] or the problems encountered with redistributing tasks between doctors and nurses, or the normativities of researchers from the field of science and technology studies (STS) engaging with the construction of care practices.[5] However, in this essay we will focus on the historic specificities of the situation in which we were asked to *develop standardised care trajectories* for a hospital that claims to *work towards patient-centred care*.

The slogan Hospital E used in a large ad campaign at the time of the research was '*heel de mens*', which is a combination of 'heal the human being' and 'the entire human being'. This merged focus on cure and on integral care seems to have gained another ally in the hospital's pathway experiment: standardisation. For a long time standardisation and patient-centred care have formed perfect opposites in the medical sociology literature. Substantive critiques by Anselm Strauss, Shizuko Fagerhaugh, Barbara Suczek and Carolyn Wiener[6] pointed out that the imported business jargon of the Standard Operating Procedure would sit uncomfortably with important aspects of care delivery they famously termed 'articulation work', while the more rhetorical standardisation critiques by authors like George Ritzer[7] indicated that patients as well as physicians are increasingly under the control of large, standardised systems. Care practitioners have a reputation for resisting standardisation, arguing that this is a problem for care delivery.[8] The invitation by clinicians to do this project therefore invokes the question of how the concepts of 'standardised care trajectories' and 'patient-centred care' relate. Which historical specificities result in the pragmatic commensurability of these concepts? How could the dichotomy between standardisation and patient-centredness be reconfigured? Which kinds of patient-centred care are enacted through the development of certain types of standardised care? With the development of standardised care pathways and their potential to bring together issues in care work and standardisation,[9] both the concepts of patient-centred care[10] and care trajectories seem in need of re-specification.

In order to address these questions, we will first situate our research practice in a tradition of ethnographic research of work practices while also indicating what is particular about such research of care practices in our present timeframe. Second, we deal with two different conceptualisations of pathway development and of patient-centredness. We will conclude by stating how these conceptualisations relate and which lessons can be drawn from the histories of pathway development for oncology patients.

## Different ethnographies

The idea that ethnographies can be used for creating better (care) practices is dominating scientific fields such as Participatory Design (PD) and Computer Supported Cooperative Work (CSCW). Drawing on the ethnomethodological work of Harold Garfinkel, many publications claim to have come to 'better' ICT applications that really support work practices, rather than support or automate idea(l)s of work practices.[11]

Though the work of Garfinkel and of authors from CSCW and PD is largely inspiring due to the strength of its detailed analyses, there seems to be an inherent conservatism in such studies that seems rather awkward in the light of the observation with which we started this chapter. These ethnographic analyses tend to get stuck in a fascination for complexity and end up with extremely modest conclusions about the way in which things could be different. A classic example of this *problematique* is Garfinkel's brilliant chapter 'Good organizational reasons for "bad" clinical records'.[12] In this chapter Garfinkel critiques the bewilderment that is displayed time and again about the poor quality of medical records. He rather investigates *why* records are 'bad': 'The sheer frequency of "bad records" and the uniform ways in which they are "bad" was enough in itself to pique our curiosity'.[13] He concludes that the troubles with medical records should be seen as:

> 'normal, natural' troubles [... which are] troubles that occur because clinic persons, as self-reporters, actively seek to act in compliance with rules of the clinic's operating procedures that for them and from their point of view are more or less taken for granted as right ways of doing things.[14]

By this he means that they are 'in accord with prevailing rules of practice (...) [and that they are] integral features of the clinic's usual ways of getting each day's work done – ways that for clinic persons are right ways'.[15] Now it is this last brief sentence that fascinates us: the conclusion that *because* it is 'common practice', it is also 'right'. This conclusion is strongly linked to Garfinkel's being taken with the relocation of normativity *in* practices – what he called 'practical ethics'[16] and leads him to conclude that in the case of medical record keeping the shortcomings are:

> persistent, they are reproduced from one clinic's files to the next, they are *standard* and occur with great uniformity as one compares reporting systems of different clinics, they are obstinate in resisting change, and above all, they have the flavor of inevitability.[17]

What the health care professional in our lecture still noted ironically has here become a serious ethnomethodological claim: the problems of medical record keeping are standardised!

This analysis has been helpful in counter-arguing the continuing claims that medical records are 'carelessly kept' – including the implied cognitive 'solutions' of 'disciplining' and 'educating' medical professionals – and that if only care workers would be more careful, records could unproblematically be used for medical research purposes.[18] We do however feel that the fascination for complexity in ethnomethodologically-inspired workplace studies makes them conservative to such an extent that they fail to adequately address the problems that patients and care professionals experience because of *ad hoc* ways of dealing with 'normal, natural troubles' in health care. By locating ethics *in* work practices as Garfinkel does, there is little attention to normativity that is also partially connected to *other* practices than the *work* practice under study, such as the work practices of quality improvement agents, nor to the fact that there are *different work practices* taking place simultaneously. 'Practical ethics' is conceptually problematic since sociologists have focused on the multiplicity of practices.[19] What about all the referral patients to Hospital E who are referred without proper information accompanying them due to poor record keeping and because of which terminal patients have to wait for a long time to have diagnostic tests done that were possibly already carried out in the referring hospital? What about the *bad* organisational reasons for bad clinical records? Rather than ethnographically blurring the problems that patients are facing in practices of care delivery, we try to use ethnography to elucidate particular possibilities for how things could be different. See for example this excerpt from an observation at the haematology/oncology treatment centre:

A lady in her late fifties enters the treatment center. She is looking skittishly around and is walking with unsure steps. It is clear to see that she is coming for the first time and does not know her way yet.

Gina, one of the nurses at this station, tells the lady that she may select a chair and explains how the adjustments function:

'It's convenient to move up the armrest and get in then. That way the chair is not so high. And you can get out like that too. Soon you will have to remain seated for your course for four hours. At the end of this room, around the corner, there is a restroom. It would be good if you would use that now, because than you don't have to walk with the drip afterwards. You will be getting Taxol in a bit. Have you had a course here before?'

'No, here not yet', the lady replies.

'I thought you had never had a course before?', Gina continues to ask.

'Well, I did ... When was that ... I think that's already one, perhaps two years ago', she replies somewhat confused.

Gina explains the steps of the treatment rapidly; first one bag, then the next, something against sickness, something against reactions to treatment, etc. – it looks like the lady can't follow nor remember it all.

'Well, good, if I'm starting to be convulsed, I'll let you know', she says jokingly.

'Well, you may be laughing now, but this can really happen', Gina says. 'You can have a reaction and you can in fact be convulsed then. But that's not very likely'.

The lady is visibly shocked, after which Gina emphasises once more that there is only a minor chance for such reactions. After this, the lady goes meekly to the toilet, comes back and the drip is installed.

'Dear ...' Gina sighs afterwards: 'previously we always had an introduction talk with someone like her. We had a small, separate room in the back for that. But even there we have installed a bed for treatment these days. We have grown so much that we are unable to deliver some very basic forms of care. When on top of that one realizes that the surgery hours are completely overbooked, which implies that there will be no time for their questions there either, I sometimes wonder what kind of care we are actually delivering here ...'

This account does not serve the purpose of merely understanding the practice of care delivery at this ward as such. It also does not aim at elucidating the skilful ways in which care workers keep a health system under stress smoothly running. It is rather inspired by a concept of other ways in which the care setting under study could be organised and how painful events like this one could be prevented.

With this aim of addressing particular problems in practices of care delivery, we are siding with many others in our present timeframe. Influential reports published by the Committee on Quality of Health Care in America of the US Institute of Medicine (IOM) have claimed that with increasing life expectancy, chronic conditions on the rise, a continuous increase in the complexity of medical treatment and the need for cost-containment, drastic improvements of Western health care systems are crucial.[20] In the Netherlands, similar voices were raised in relation to the present and potential performance of the health care system.[21] Simultaneously, colleagues from the fields of STS and feminist technology studies have been voicing what we would like to call a 'critique of critique' and are propagating more engaged and co-constructive approaches to the development of (health care) practices.[22]

Garfinkel had good reasons to claim that it was 'the least interesting thing' to say that problems with medical record keeping were due to careless data entry by care professionals.[23] However, in the light of the developments we relate our study to, it now seems in a way an equally uninteresting thing to get stuck in a glorification of existing practices while knowing the costs patients and care professionals sometimes have to bear.

However, rather than seeing standardisation as the solution to all problems, as is championed by groups like the IOM's Committee on the Quality of Health Care in America, we prefer to take concrete health settings and the problems we encounter within them as a starting point. Siding with Steve Brown who stated that 'it is quite easy to come up with solutions; these are merely implied in the way in which we define problems',[24] we realise that how we were involved in developing standardised pathways as a solution – though in a more general sense present from the start of our study – will always depend strongly on the particular problems encountered in the practice under study. Though in a way standardisation may be a solution in search of a problem, the particular form of this solution is being strongly shaped by the ethnographic analysis of the practice in which it is developed. This means the *form of standardised care trajectories* is linked to the analysis of care practices, as is the form of patient-centredness we hoped to enact through these pathways. This makes standardisation and patient-centred care situated notions that co-construct each other.

## Dissecting standardised care trajectories

In order to explicate which forms of standardisation and patient-centredness can be made commensurable, it is important to differentiate between various ways of developing standardised care trajectories. We would therefore like to dissect both concepts.

At times when the health care system is being displayed as under substantial stress, the advocates of the panacea of un-situated standardisation continuously emphasise the high rates of medically inexplicable practice variation,[25] the importance of privileging data over judgement and the emancipatory possibilities for patients and other care providers to make decisions in cooperation with their clinicians if only the latter would work in accord with guidelines.[26] The strong claims by proponents that standardisation of health care practices can provide impressive increases in efficiency,[27] effectiveness[28] and patient and professional satisfaction,[29] make it appear like the knight in shining armour of a medical system under stress – just as standardisation was seen as the saviour for US business firms warding off the successes of their Japanese competitors in the 1980s.[30]

The Integrated Care Pathway organisation especially seems to thrive on this rhetoric and reifies the standardisation of care. Since the solution they take as a starting point is the development of Integrated Care Pathways, they focus mainly on the methods needed for creating pathways and on ICT that supports their implementation. In Belgium and the Netherlands the Clinical Pathways Network has developed a 32-step methodology that includes writing out each step a patient goes through in a time/task matrix, setting up Service Level Agreements with all supporting units, training personnel and finalising a procedure determining who is responsible for the entire management of the clinical pathway[31] *prior to the implementation phase.* Standardisation is conceptualised here as a *solution per se* that is being *developed at considerable distance* from the particular problems of work practices and that needs to be *implemented after all standardisation has been done on paper.*

It is this kind of standardisation that critics refer to as 'assembly-line medicine'[32] of which it is feared that simple, economically interesting care trajectories are standardised to a great extent at the expense of other forms of care. According to Ritzer, the foremost critic of standardisation in any form, whose classic work *The McDonaldization of Society* has recently seen its sixth edition, this development in health care will ultimately lead to the 'dehumanization and depersonalization of medical practice'[33] and is part merely fitting the drive for efficiency that he so whole-heartedly fights in his McDonaldization thesis.

In order to overcome the analytical gridlock that this polarised debate inflicts, Stefan Timmermans and Marc Berg have stressed the need to study standardisation in health care in a less antagonistic way. They 'propose a study of the *politics of standardization in practice*',[34] that focuses on the actual changes in medical practice as a result of standardisation, on the perceivable renegotiations of orders and autonomies that come with the standards and on how certain standards and universalities actually came about. Taking as a starting point that '*standardization is*, paradoxically, *a dynamic process of change*'[35] in which different standards order worlds in different ways, they open up the analytical grounds that were closed by the vigorous controversies. What Timmermans and Berg, however, merely hint at, rather than empirically explore, is the space their move equally opens up for situating the activity of standardising care practices and 'do politics through standardisation'.[36] It is this more situated standardisation that we propose as an approach that may well align with particular forms of patient-centredness and may not necessarily undo the potential for articulation work emphasised by Strauss *et al.*[37]

In order to situate standardisation in the specific problems of care practices and relate them to a conceptualisation of what the organisation of care at this ward could be like, it is of course crucial to get an understanding of them. Therefore we combined ethnographic approaches such as participant observation (19 days), semi-structured interviews with resident staff, junior doctors, operational management, research nurses, medical secretaries and medical social workers (23 interviews in total), focus-group project meetings (twice), and interactive presentations to nurses (twice), haematologists, oncologists and other personnel of the clinic (twice) with the analysis of data from the hospital information system. Quantitatively, this allowed us to analyse to what extent the clinics were running late, how many patients were booked in the agendas for whom there was not a regular slot available (double bookings or over bookings), whether clinics started on time, whether there was an overall balance between the capacity of the clinics and the number of patients that visited, how many cancellations of clinics by doctors took place and the amount of variability on these parameters for individual doctors. We also quantified the increase in the number of treatments given at the treatment centre over the last years and the distribution of treatments over the various days of the week over the last three months. We did not look at the 'no show' indicator that is often relevant to improve capacity use at overcrowded wards, since the patients at this ward are so ill that they hardly ever stay away without notification. We also involved ourselves in discussions and publications on quality of care and health care innovation and their critiques from medical sociology and STS, for example by presenting our work at the seminar that formed the starting point of this collected volume.

This approach was not part of a 'design phase' in which the solutions were thought out, but it was a way to become engaged with the practice of care delivery at the ward and with potentially promising practices of health care innovation in this setting. It proved most fruitful for finding spaces to *act with* various actants in the ward, such as doctors, the planning module of the hospital information system, nurses, the hospital IT department, doctor's assistants, strategies for working together with hospitals in the region, secretaries, plans for renovating the ward, cluster management, the finance and control department, etc.

After this first phase, we wrote a proposal for possible changes for this ward. We concluded that there were three problems that the ward was facing: the strong variability of workload during the week, the absence of a planning system for the treatment centre of the ward, and the overcrowded surgery hours of both haematologists and oncologists. Some implied solutions therefore were: reducing variability by shifting surgery hours around to find a better distribution over the week; developing a planning system; and creating space in the surgery hours of the doctors. When this analysis and its directions for a solution were accepted, the medical coordinators[38] were positioned 'in the lead' of the project by making them chairs of the working groups for haematology and oncology.

These multi-professional groups, consisting of the medical coordinators, staff members, nurses, doctor's assistants, research nurses, the management of the secretaries, the operational manager of the ward and one of the authors (TZJ) of this chapter, met once a week to discuss the progress of the project and to work out care trajectories for the large majority of patients. Since the haematologists were keeping a diagnostic registration, we could easily assess that we would need 12 standardised care paths[39] to cover 69 per cent of the patients, which we decided would be a large enough group to have a serious impact on the work of the ward. Since

*Figure 10.1*   Care trajectory flow chart for Hodgkin's lymphoma

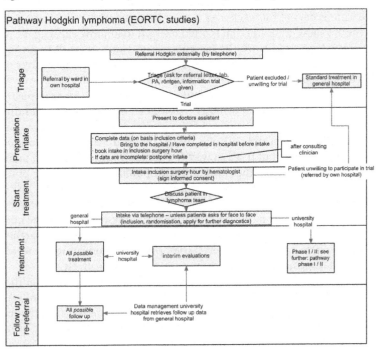

the oncologists did not keep such a registration, we had to make an estimate for the number of care paths that were needed to cover approximately 80 per cent. This turned out to be possible by developing six trajectories.[40] Part of the discussion about how these trajectories should be organised centred on drawing up sketches of flow charts for each diagnosis, such as those shown in Figures 10.1 and 10.2.

The set-up of multi-professional working groups, with the medical coordinators as chairs, made our role one of 'acting with', which meant somehow giving up the freedom to propose interventions or get them accepted, but also distributing ownership of the project and its interventions. It was also an extra assurance that proposed solutions would correspond with particular needs of the ward's professionals.

We tried to stress the importance of seeing the development of care trajectories as a way to change health care organisations from their unit-based organisational structure (outpatient clinic, laboratories, radiology

*Figure 10.2*   Care trajectory flow chart for oesophageal carcinoma

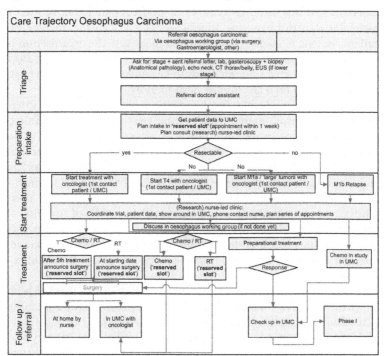

department, operating theatre, clinical departments, and so on) to a process-based organisation focused on the trajectories that groups of patients go through.[41] This means that we questioned the functional organisation of hospital care with its internal inconsistencies, and brought the trajectory that certain groups of patients follow to centre stage.

The conceptual shift made by this approach to standardisation is that it moves away from the standardisation of trajectories of individual patients or of the process of decision-making by individual specialists. Rather than trying to standardise processes that have been shown by STS researchers,[42] medical sociologists[43] and ethnomethodologists[44] to be extremely complex, dynamic and unpredictable, standardisation is focused on an *aggregated level*, i.e. the level of *patient groups*. At this level it turns out that trajectories display substantial similarities. Whereas individual patient trajectories are inherently unpredictable, the trajectory for groups of patients can be *made* predictable and care can be organised accordingly without running into the problems of frustrating articulation work at the level of individual patients. Even if certain steps cannot be planned for specific patients due to the inherent variability of individual care trajectories, it is possible to assess for a group of patients how often a particular step (for example an emergency CT scan for oncological patients having a relapse) will occur and make sure the organisation is ready to meet this demand (for example by keeping emergency slots available until such a CT scan needs to be made). These capacities can then be assigned on a last minute basis to individual patients.

In this approach, developing standardised care trajectories is inter-related with four domains: re-delegating tasks, integrated planning, performance management and process-supporting ICT (Figure 10.3). These domains are to some extent the 'usual suspects' in standardisation initiatives. Re-delegating tasks, is for example, one of the classical promises of 'manualised' treatment.[45] In this line of reasoning, treatment is standardised first and the re-delegation of tasks carried out afterwards as a way to implement and maximise the results of the clinical guideline. When situating standardisation, re-delegating tasks is part of the dynamic process of standardisation itself: those aspects where re-delegation seems professionally feasible shape the form of the standardised care trajectory.

Similarly, performance management is a worn-out trope that has been so heavily critiqued[46] that we need to be specific as to what would make it 'situated' here. Rather than the usual norm-setting activity of defining what the optimal medical care is at a distance from care delivery and deciding which indicators should be used to manage the treatment of

*Figure 10.3* Domains of situated standardisation. Originally published in a slightly different form in Berg *et al.* 2005 and used with permission of the authors.

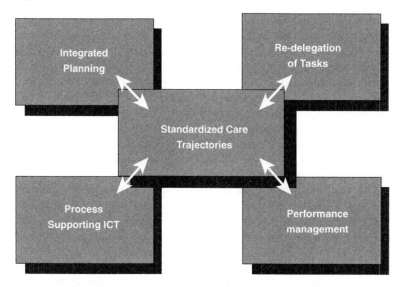

particular groups of patients, the indicators are co-constructed with the standardised care trajectories for a particular care setting and interwoven with management information and common and highly local management practices.[47] Integrated planning may be seen as the rationalisation of the steps patients go through that is then implemented through organisational changes, but in this empirical quest of situating standardisation it proved to be the organisationally opportune capacities that defined which parts of the care trajectories were planned integrally. Finally, process-supporting ICT is, of course, often seen as the ultimate tool for the coordination of highly complex standardised care processes.[48] However, in this approach of situated standardisation, rather than developing overarching ICT applications which coordinate entire care trajectories – and running into 'implementation problems' at a later stage – major gains in the manageability of care trajectories were realised by aligning existing modules of the hospital information system with the aims of solving, for example, the persistent problems of planning at the treatment centre of the outpatient clinic.

When focusing on the way these five domains are intertwined, the standardised care trajectories are no longer the *point of departure* for an organisational change and implementation phase that follows the design phase. They are the *outcome* of the project as a whole: it was only apparent how the care trajectories would look when we explored which organisational interventions were feasible, an approach we have elsewhere called 'preventing implementation'.[49] Since the *organisational* problems rather than the problems faced for a particular diagnostic group, remained the starting point of the project, there was also no issue of developing a standardised care trajectory that would create problems for other types of patients. This is an important difference from the approach of the Integrated Care Pathways movement, which tends to focus on pathways for a particular diagnostic group, such as hip replacement, but then risks improving the organisation of this care trajectory at the expense of other, perhaps less common types of diseases. When at this ward it was for example mentioned by both doctors and patients that their biggest frustration was the absence of data from the hospital that referred the patients during their surgery hours at Hospital E, and that patients needed many visits before treatment would start because of suboptimal referrals and preparation of the intake visit, a doctors' assistant was installed to make sure that all relevant diagnostics would be done and the results would be available before the intake visit with the specialist in Hospital E. This preparation of the intake was supported by a newly developed ICT module that was part of the development trajectory of the electronic medical record (EMR) at hospital E and saved time and frustrating encounters for both doctors and patients. The first step of *all the standardised care trajectories* therefore became a thorough preparation of the intake. This intervention was carried out for all patients who were scheduled for a new appointment, which means this was also a relevant improvement for patients who did not belong to any of the 18 care trajectories that were developed in this project.

## Dissecting patient-centredness

Just now we mentioned that we heard about the frustrations of patients at this ward. We met them during the surgery hours we observed and afterwards saw them in the waiting room. We heard what they said at the counter when making their appointments. We even met one of their 'veterans', who had been treated there for more than ten years, at a birthday party. Interestingly, this common friend was an organisational consultant working for a large consultancy firm. One of us (TZJ) had a number of private meetings with him to check proposed changes and get input from this 'expert patient'. However, a more structured way of involving patients was not a formal part of the setup of this project. One of the reasons for this was that doctors seemed highly sceptical about what the involvement of patients would add. We feel that the 'nothing about me without me' rhetoric that is so warmly embraced by health care improvement institutes around the world (e.g. the Institute for Healthcare Improvement in Boston), does more harm than good in this regard. Patients who are presented as increasingly demanding customers who are always right, seem highly popular in policy documents on market developments in health care.[50] Clinicians, however, have valid reasons to be suspicious about such assumed consumer-patients. This form of patient-centredness individualises patients, stressing their uniqueness, and draws rhetorical strength from claiming to be 'what care really is about'. It is also the kind of patient-centredness that is often used to rebuke hospital management or interventionist researchers about the way they overlook the specificity of the trajectory of individual patients or to criticise quality of care researchers for what Ritzer would call their contribution to the McDonaldization of society.

'Consumers' who demand things like one-stop visits since they are 'accustomed to efficiently organised McDonaldized systems'[51] may – according to Ritzer – actually pay a very high price for their demands:

> The drive for efficiency can make them feel like products on a medical assembly line. The effort to increase predictability will likely lead patients to lose personal relationships with physicians and other health professionals because rules and regulations lead physicians to treat all patients in essentially the same way. This is also true in hospitals, where instead of seeing the same nurse regularly, a patient may see many different nurses. The result, of course, is that such nurses never come to know their patients as individuals. As a result of the emphasis on calculability, the patient is more likely to feel like a number in the system rather than a person. Minimizing time and maximizing profits may lead to a decline in the quality of health care provided to patients … Thus, the rationalization of medicine has increased the dehumanization and depersonalization of medical practice.[52]

But besides the theoretical critique of this conceptualisation of the patient as customer, there were also reasons from within the ward that made us question the centralisation of individual patients as the right way to go forward.

There was one oncologist who seemed to define her work as treating customers who were always right. She was notorious for allowing her surgery hours to run over time and she had indicated from the start that she had no time for an interview with us. She rejected our invitation, stating she was spending her precious time on patient care. When one of us reported this to the medical manager of oncology he got really irritated, walked into the secretariat and summoned the secretary to make an appointment for this unwilling oncologist for the interview.

Though she was late, she came for the interview. However, within a few sentences it became apparent that she would not be cooperating readily. When we introduced the project and the aims of improving the organisation of care at the haematology/oncology ward she snappishly replied:

'And 'heel de mens',[53] where does that fit in?!'
TZJ: 'You're still missing that?'
'I don't know, I don't see it'
TZJ: 'What do you mean, you don't see it?'
'Well, where do you see it?!'
TZJ: 'Well, where you should start seeing it, in this project is when patients who now have to wait for a very long time to get their diagnostic trajectory organized will be helped in a much smoother way ...'

A skilled interviewee: rather than allowing the interviewer to determine the direction of the conversation, she takes over the interview and forces the interviewer to demonstrate his good intentions. However, his explanations were obviously not convincing. Soon she interrupted again:

One more thing, what will happen with those tapes? Who will have access to them?

We later found out that her position was under serious threat and that her suspicion was grounded. Doctors' assistants at the counter had problems with her clinics because she was always running late and this was hindering the flow of the work. When observing one of her surgery hours it became apparent why consultations with her took this much time: she was willing to discuss any problem that a patient encountered and defined nothing as beyond her scope. When patients had difficulties deciding whether or not to book a holiday because they did not know whether their course of treatment would be postponed, she would go into the details of explaining that a last-minute booking would perhaps be most suitable, and explain which websites were offering such holidays. Though this seemed to be the most amazing service that an oncologist could provide to her patients, it meant patients had to wait long hours in the waiting room. Often patients who were scheduled later in the morning or afternoon called in advance to hear how much she was behind schedule. Her 'patient-centred care' also meant doctors' assistants could not go for lunch, would have extra work during the afternoon sessions that had piled up during the morning or would not be able to go home in time. If they did go home before the clinic was over, a stack of records of patients needing appointments would be left behind for the morning crew who would have to call those patients.

Ironically, this reification of the individual wishes of patients at times seemed to shift into a regime in which the oncologist completely controlled patients. Nurses from the treatment centre expressed some doubts about her medical treatment and found it hard to treat patients that were under her care. At one time, one of them shared with one of us that she couldn't face the 'stress schemes' this oncologist was working with any longer. In oncology trials, these schemes indicate what should be done when a patient has a reaction to the treatment, meaning their bodies no longer are able to take the often quite aggressive chemotherapy treatment. The treatments at this ward were generally subjected to rigid trial protocols and it was clearly indicated after how many reactions a patient had to stop treatment, which, in the case of these tertiary referrals, often would mean their care trajectory would become a purely palliative one. However, this oncologist had her own additional stress schemes for the instances in which a patient was having too many reactions to be allowed to continue treatment. The main exception was that she introduced extra breaks of about half an hour after which the course was re-started. This was after the patients already had the number of 'time-outs' that were approved by the protocol. This nurse indicated that she felt she was forcing patients through treatment that was no longer bearable for them and pointed to the lack of attention to the fallibility of treatment and palliative care.

Medical management was aware of these problems and had intervened in their own ways. This oncologist was given a strict maximum of seven patients she could see per surgery hour – while others had 12 slots and were actually seeing up to 18 patients to make up for the extra time she was taking for her patients and prevent an increased access time to the surgery hours. As a consequence, those patients that did not fit her schedule any longer were booked with other doctors. This was highly inefficient, according to her, and she commented on this during the interview:

> We have to be able to treat our patients properly, our own patients, and not *ad hoc* from one to the next doctor which makes you lose grip and enables things to just go wrong. Those things are not in the interest of patients, nor of us. It seems like you're being relieved but it gives twice as much work. Because this other doctor sees the patient once between things, but he does not have the story and the policy in his head. And sometimes things are not done in a proper way then, or you miss certain things. It is just not in the interest of the course. And afterwards I get all these things to be arranged on my desk, all the phone calls of things that are not correct. So finally, if you just would have taken a bit more time during the surgery hour, you would have finished much earlier.

When indicating the possibilities for having a nurse practitioner take over substantial parts of the treatment and improving the cooperation with this professional and with other colleagues through re-installing a weekly oncology meeting she stated:

> Well, but of course it is ... we can do it without a doctor. I see people from the periphery[54] who say: 'I've seen the doctor at the beginning and at the end of the chemotherapy and besides that I haven't seen him'. So it's possible. I guess it can work out fine with all these nurse practitioners, but I'm used to intervening a bit and treatment can easily be adjusted to small complaints. And that's why I see people briefly [before each time they get treatment] ... And an oncology meeting doesn't cover everything and I think it shouldn't: that would only take more time. I thought we were working on solutions?! To improve things?! ...

In the end of the day she successfully resisted attempts to change her 'patient-centred care'. Patients found out that she was having less time for them and that they were often referred to colleagues. Some of them apparently told her that they were surprised about this change of policy. She then encouraged them to file an official complaint to the medical professor in charge of the ward, since they were no longer receiving the quality of care they experienced before. As a result this professor spent long hours answering those complaints and decided that the imposed sanctions on her should be lifted.

Of course this is a rather extreme case of how patient-centredness at the level of individual patients is not merely problematic, it also at times excessively empowers care professionals allowing them to use patients and their ostensible wishes to justify organising their work in the way they prefer. Within a health care setting, 'playing the patient card' is a powerful tool. When patient trajectories are unique and patients should be treated by the organisation as individuals, all kinds of inter-professional checks are bound to be weak. An oncology meeting at which the use of the scarce time for surgery hours, or adjustments in stress schemes would be discussed, would replace the individualised notion of patient-centred care this oncologist was holding on to a notion of patient-centredness in which patients were seen as groups over which scarce resources needed to be distributed, and oncologists were seen as a collective jointly reflecting on issues such as deviations from stress schemes. This indicates that by focusing primarily on patient-centred care as situated in the interpersonal encounter between two human beings – a doctor and a patient – other possible issues of delivering highly complex patient-centred care are easily overlooked. What may at first sight look very sympathetic and patient-centred can in practice turn out to be highly coercive.[55] The risk of an individualised notion of 'patient-centred care' empowering doctors in unprecedented ways is a phenomenon that is insufficiently documented in the medical sociological literature.

The kind of patient-centred care that we feel is *not* at odds with situated standardisation of care, conceptualises patients as part of a collective. From the perspective of a health care organisation, patients are part of a population that makes use of the same capacities, such as outpatient clinic space, doctors' time, radio-diagnostic capacity, clinical beds and chairs in the treatment centre. If patients are treated as individuals who all should receive individually optimal treatment, without taking the collective that they are part of into account, the organisation will pay a high price – as may various individual patients at times. And even if patients are seen as a collective, there may be other collectives that need to be taken into account for care to be patient-centred in practice:

> When I arrive at the treatment center at around 11 am I hear the nurses have shifted two patients who would have come today. A few days before they saw this coming and then one of them has called two patients to check if they could be moved to another day.
>
> 'If people come for a course I don't shift them around that easily but these people needed something else. An additional infusion of

calcium for example. And if I see on Friday that next week will be too full, I call them with a lie for their own good. Something like: 'the doctor is on holiday' or 'at a conference' or something, but that the next day he will be back. Of course I'd rather not do that and some understand it very well and to those I just explain what the situation is. But of course we know the people here so we also know who will be whining ...'.

Actually the problem is that the treatment center is overcrowded at certain times. There seems to be a peak load on certain days. When we try to figure out what went wrong in this instance, it turns out that hematology and oncology patients are being scheduled on the treatment center without those two specialties being aware of what the other is booking. They both plan for their own patients while using the same chairs. This often leads to such late shifting around of patients or to patients getting their treatment while sitting on a stool rather than on an adjustable chair – with all the risks that implies.[56]

It therefore seems crucial to not merely see patients as part of a certain diagnostic group[57] but as part of a collective that is defined by the resources that are being used. This means that *patients are part of different collectives at different times*. Defining patient-centred care as a mode of organising that prepares the organisation for the various collectives of patients it serves, also enables this organisation to distinguish collectives that have different requirements.

## Conclusion: The patient-centred care trajectory and its lessons

On the basis of the project in Hospital E we have dissected both 'patient-centredness' and 'standardisation', the classical opposites in the medical sociological literature. What we hope to have shown is that the reconceptualisations we have proposed offer a way out of the analytical gridlock that proponents and critics of standardisation tend to get stuck in.

The ways in which patient-centred care and standardisation are being brought together in care pathway development calls for more sociological analysis, but also for an enhanced sensitivity to the lessons that can be learned from previous attempts to standardise care 'in the interest of patients'. Our research suggests that definitions of patient-centred care as serving customers who are always right and of developing standardised care pathways as an unproblematic solution, are equally harmful to the actual problems patients and professionals face in contemporary care practices. Given this conclusion, much of the work in the field of health services research that privileges standardisation over local complexity, is in serious need of sociological sensitivity to the ways in which articulation work can be made part of standardisation initiatives. Similarly, the critics of standardisation initiatives need to offer us more specific analyses of patient-centred care and standardisation. Otherwise they risk praising forms of patient-centredness that promise to serve individual patients, whereas these are actually a disservice to patient collectives by placing excessive power in the hands of medical professionals.

## Notes

1  The names in the excerpts have been changed.
2  Cf. Zuiderent-Jerak 2007; Zuiderent-Jerak 2009; and Zuiderent-Jerak and Berg 2010.
3  Zuiderent 2002; Bal and Mastboom 2007.
4  Dramatic decreases in the time clinics were running late while at the same time increasing the production of the ward.
5  Zuiderent-Jerak and Jensen 2007; Zuiderent-Jerak 2010.
6  Strauss *et al.* 1997.
7  Ritzer 2011.
8  Cabana *et al.* 1999.
9  Allen 2010.
10  Stewart *et al.* 1995; Mead and Bower 2000.
11  Garfinkel 1967; Blomberg *et al.* 1996; Procter and Williams 1996; Button 2000; Luff *et al.* 2000; Suchman 2000.
12  Garfinkel 1967.
13  Garfinkel 1967, p. 191.
14  *Ibid.*
15  *Ibid.*
16  Garfinkel 1967, p. 74.
17  Garfinkel 1967, pp. 191–2, emphasis added.
18  Garfinkel 1967, p. 192. This contribution has proven invaluable in the last decades to argue against the development of electronic medical records that provided new and unprecedented opportunities for disciplining the use of the medical record by care professionals; see Berg 1997.
19  Mol 2002; Zuiderent-Jerak 2010.
20  Kohn *et al.* (eds) 2000; Committee on Quality of Health Care in America 2001.
21  Schrijvers *et al.* 2002.
22  See e.g. Berg 1998; Kember 2003 and Latour 2004.
23  Garfinkel 1967, p. 192.

24  Discussion with the author at 'Does STS mean Business I' conference, June 2004, Said Business School, Oxford University.
25  Wennberg and Gittelsohn 1973; Wennberg 1984.
26  Woolf *et al.* 1999.
27  Evans and Hwang 1997.
28  Berdick and Humphries 1994.
29  Ford and Fottler 2000.
30  Hammer 1990.
31  Vanhaecht and Sermeus 2002.
32  Ritzer 1992, p. 43.
33  *Ibid.*
34  Timmermans and Berg 2003, p. 21, emphasis in the original.
35  Timmermans and Berg 2003, p. 23, italics in the original.
36  Timmermans and Berg 2003, p. 216.
37  Strauss *et al.* 1997.
38  These are doctors who are responsible for managing the medical part of the work on the ward. They are the medical part of the dual management, the 'organisational' part being covered by an operational manager.
39  Hodgkin lymphoma, non-Hodgkin lymphoma < 65, non-Hodgkin lymphoma > 65, Recurring follicular non-Hodgkin lymphoma, chronic lymphatic leukaemia general, chronic lymphatic leukaemia Campath, multiple myeloma < 65, multiple myeloma > 65, haemostasis, thrombosis, haemophilia, Phase I/II trials.
40  Testicular carcinoma, bladder carcinoma, colon carcinoma, oesophageal carcinoma, ovarian carcinoma and Phase I trials.
41  For a further elaboration of this approach, see Berg *et al.* 2005.
42  Berg 1997.
43  Wiener 2000.
44  Garfinkel 1967.
45  Hayes *et al.* 1999.
46  See for example Porter 1997 and Power 1997.
47  For an interesting example of how such 'grown' indicators systems can become strong governance instruments, see Beersen *et al.* 2005.
48  See e.g. Sermeus *et al.* 1997.
49  Zuiderent-Jerak 2007.
50  See e.g. RVZ 2005.
51  Ritzer 1992, p. 45.
52  Ritzer 1992, p. 138.
53  The previously mentioned hospital slogan.
54  University hospital jargon for 'general hospital'.
55  Silverman 1987.
56  The main reason for having treatment chairs is that patients who do not respond well to treatment can be cared for and their seat can be adjusted. They also are protected from falling in case of reactions to treatment if they are in such a chair. Treating patients on a stool is a serious violation of the basics of patient safety in oncology treatment.
57  Which is strongly encouraged by developments in the financial structure of many Western health care systems towards payment according to Diagnosis Related Group (DRG) classifications.

# Works Cited

Allen D. 2010, 'Care Pathways: Some Social Scientific Observations on the Field', *International Journal of Care Pathways*, 14, 47–51.

Bal R. and Mastboom F. 2007, 'Engaging with Technologies in Practice: Travelling the North-West Passage', *Science as Culture*, 16, 253–66.

Beersen N., Redekop W. K., de Briujn J. H., Theuvenet P. J., Berg M. and Klazinga N. S. 2005, 'Quality Based Social Insurance Coverage and Payment of the Application of a High Cost Medical Therapy: The Case of Spinal Cord Stimulation for Chronic Non-Oncologic Pain in The Netherlands', *Health Policy*, 71, 107–15.

Berdick E. and Humphries V. 1994, 'Hospital Re-engineers to Improve Patient Care', *Health Care Strategic Management*, 12, 13–14.

Berg M. 1997, *Rationalizing Medical Work: Decision-Support Techniques and Medical Practices*, Cambridge: MIT Press.

Berg M. 1998, 'The Politics of Technology: On Bringing Social Theory into Technological Design', *Science, Technology, & Human Values*, 23, 456–90.

Berg M., Schellekens W. and Bergen C. 2005, 'Bridging the Quality Chasm: Integrating Professional and Organizational Approaches to Quality', *International Journal of Quality in Health Care*, 17, 75–82.

Blomberg J., Suchman L. and Trigg R. 1996, 'Reflections on a Work-Oriented Design Project', *Human-Computer Interaction*, 11, 237–65.

Button G. 2000, 'The Ethnographic Tradition and Design', *Design Studies*, 21, 319–32.

Cabana M. D., Rand C. S., Powe N. R., Wu A. W., Wilson M. H., Abboud P-A. C. and Rubin H. R. 1999, 'Why Don't Physicians Follow Clinical Practice Guidelines? A Framework for Improvement', *Journal of the American Medical Association*, 282, 1458–65.

Committee on Quality of Health Care in America, Institute of Medicine 2001, *Crossing the Quality Chasm: A New Health System for the 21st Century*, Washington: National Academy Press.

Evans J. H., Hwang Y. and Nagarajan N. J. 1997, 'Cost Reduction and Process Reengineering in Hospitals', *Journal of Cost Management*, 11, 20–7.

Ford R. and Fottler M. 2000, 'Creating Customer-Focused Health Care Organizations', *Health Care Management Review*, 25, 18–33.

Garfinkel H. 1967, *Studies in Ethnomethodology*, Englewood-Cliffs, NY: Prentice-Hall.

Hammer M. 1990, 'Reengineering Work: Don't Automate, Obliterate', *Harvard Business Review*, 69, 104–12.

Hayes S. C., Barlow D. H. and Nelson-Gray R. O. 1999, *The Scientist-Practitioner: Research and Accountability in the Age of Managed Care*, Boston: Allyn and Bacon.

Kember S. 2003, *Cyberfeminism and Artificial Life*, London: Routledge.

Kohn L. T., Corrigan J. M. and Donaldson M. S. (eds) for Committee on Quality of Health Care in America, Institute of Medicine 2000, *To Err is Human: Building a Safer Health System*, Washington: National Academy Press.

Latour B. 2004, 'Why Has Critique Run Out of Steam? From Matters of Fact to Matters of Concern', *Critical Inquiry*, 30, 225–48.

Luff P., Hindmarsh J. and Heath C. 2000, *Workplace Studies: Recovering Work Practice and Informing System Design*, Cambridge: Cambridge University Press.

Mead N. and Bower P. 2000, 'Patient-Centredness: A Conceptual Framework and a Review of the Empirical Literature', *Social Science and Medicine*, 51, 1087–110.

Mol A. 2002, *The Body Multiple: Ontology in Medical Practice*, Durham: Duke University Press.

Porter T. 1997, 'The Management of Society by Numbers', in J. Krige and D. Pestre (eds) *Science in the Twentieth Century*, Amsterdam: Harwood Academic Publishers, 97–110.

Power M. 1997, *The Audit Society: Rituals of Verification*, Oxford: Oxford University Press.

Procter R. and Williams R. 1996, 'Beyond Design: Social Learning and Computer-Supported Cooperative Work – Some Lessons from Innovation Studies', in D. Shapiro, R. Traunmüller and M. J. Tauber (eds) *The Design of Computer Supported Cooperative Work and Groupware Systems*, Amsterdam: Elsevier Science, 445–64.

Ritzer G. 1992, *The McDonaldization of Society: An Investigation into the Changing Character of Contemporary Social Life*, Thousand Oaks, CA: Pine Forge.

Ritzer G. 2011, *The McDonaldization of Society*, 6th edition, London: Sage.

Raad voor de Volksgezondheid en Zorg (RVZ) 2005, *Van Weten naar Doen*, Zoetermeer: Raad voor de Volksgezondheid en Zorg.

Schrijvers G., Oudendijk N., De Vries P. and Hageman M. (eds) 2002, *Moderne Patiëntenzorg in Nederland: Van Kennis naar Actie*, Maarssen: Elsevier Gezondheids-Zorg.

Sermeus W., Hoy D., Jodrell N., Hyslop A., Gypen T., Kinnenen J., Mantas J., Delesie L., Tansley J. and Hofdijk J. 1997, 'The WISECARE Project and the Impact of Information Technology on Nursing Knowledge', *Studies in Health Technology and Informatics*, 46, 176–81.

Silverman D. 1987, *Communication and Medical Practice: Social Relations in the Clinic*, London: Sage.

Stewart M., Brown J. B., Weston W. W., McWhinney I. R., McWilliam C. L. and Freeman T. R. 1995, *Patient-Centred Medicine: Transforming the Clinical Method*, London: Sage.

Strauss A., Fagerhaugh S., Suczek B. and Wiener C. 1997, *Social Organization of Medical Work*, New Brunswick and London: Transaction.

Suchman L. 2000, 'Making a Case: "Knowledge" and "Routine" Work in Knowledge Production', in P. Luff, J. Hindmarsh and C. Heath (eds) *Workplace Studies: Recovering Work Practice and Informing System Design*, Cambridge: Cambridge University Press, 29–45.

Timmermans S. and Berg M. 2003, *The Gold Standard: An Exploration of Evidence-Based Medicine and Standardization in Health Care*, Philadelphia: Temple University Press.

Vanhaecht K. and Sermeus W. 2002, 'Draaiboek voor de Ontwikkeling, Implementatie en Evaluatie van een Klinisch Pad: 30 Stappenplan van het Netwerk Klinische Paden', *Acta Hospitalis*, 3, 13–27.

Wennberg J. 1984, 'Dealing with Medical Practice Variations: A Proposal for Action', *Health Affairs*, 3, 6–32.

Wennberg J. and Gittelsohn A. 1973, 'Small Area Variations in Health Care Delivery', *Science*, 183, 1102.

Wiener C. 2000, *The Elusive Quest. Accountability in Hospitals*, New York: Aldine de Gruyter.

Woolf S., Grol R., Hutchinson A., Essles M. and Grimshaw J. 1999, 'Potential Benefits, Limitations, and Harms of Clinical Guidelines', *British Medical Journal*, 318, 527–30.

Zuiderent T. 2002, 'Blurring the Center: On the Politics of Ethnography', *Scandinavian Journal of Information Systems*, 14, 59–78.

Zuiderent-Jerak T. 2007, 'Preventing Implementation: Experimental Interventions with Standardization in Healthcare', *Science as Culture*, 16, 311–29.

Zuiderent-Jerak T. 2009, 'Competition in the Wild: Configuring Healthcare Markets', *Social Studies of Science*, 39, 765–92.

Zuiderent-Jerak T. 2010, 'Embodied Interventions – Interventions on Bodies: Experiments in Practices of Science and Technology Studies and Hemophilia Care', *Science, Technology, and Human Values*, 35, 677–710.

Zuiderent-Jerak T. and Berg M. 2010, 'The Sociology of Quality and Safety in Healthcare: Studying a Movement and Moving Sociology', in C. Bird, P. Conrad, A. M. Fremont and S. Timmermans (eds) *The Handbook of Medical Sociology*, 6th edition, Nashville: Vanderbilt University Press.

Zuiderent-Jerak T. and Jensen C. 2007, 'Unpacking "Intervention" in Science and Technology Studies', *Science as Culture*, 16, 227–35.

# 11

## Radicalism, Neoliberalism and Biographical Medicine: Constructions of English Patients and Patient Histories Around 1980 and Now

*John Pickstone*

This chapter draws on the history of cancer, but it is mostly about patients in general. It explores how and why the possibilities of writing histories about patients have changed over the last 40 years. The first section depicts the rise of patients' history in England around 1975–85, along with then current views of patients in the NHS. The second summarises how the NHS has changed since the 1980s, notably in terms of consumerism and ideologies of choice; it then surveys developments in historiography over the same period, especially as concerns constructions of patients. The chapter ends by considering the present possibilities of writing history *for* patients – who may now be both 'expert' and 'empowered', even within public medical services.

My main suggestion is that the notion of 'bedside medicine', used in the historiography of medicine from the 1970s, can now be developed, as 'biographical medicine' – to retell illness from the patient's perspective whilst accommodating the more technical aspects of medicine as well as the more personal or cultural. This approach, moreover, is not just a device for analysts of medicine, whether in the past or present. In as much as (most) medicine is a normative activity meant to serve the interests of patients, the biographical model affords standards against which most practices of medicine can be judged.

To underline that ideal, of course, is not to neglect additional or alternative standards such as technical efficacy, equity or economy. Nor is it to pretend that most patients have (or even want?) a high degree of knowledge and agency, or that we can easily recover their agency in past times. As most of the papers in this collection show, where we do know something of patients it is often from the viewpoint of doctors for whom the agency of patients was not conspicuous. But in some of the papers, especially those that deal with very recent history, or with present-day services, we can see individual or collective agency – because public services promote it, or claim to, and because the same political shifts encourage historians to record it.

Though based on English sources, this history of 'constructions' of patients may thus be of international interest, not least because of the successive and substantial changes of structure and ideology which have been manifest recently in the English health services, albeit within a system which has so far remained tax-funded and free at the point of need. I try to show how English health politics and medical historiography have been mutually constitutive, shifting together across perhaps the major ideological reconfiguring of the twentieth century. Thus we uncover a history of medical historians as well as of patients.

## Part one: Finding patients in history around 1980

Patients' history, as exemplified in the work of Roy and Dorothy Porter, and as promoted in Roy's paper of 1982, was integral to the new discipline of social history of medicine as created in Britain from the 1970s.[1] That project involved exponents of 'social sciences in medicine' together with professional historians, notably historians of science who were keen to focus on social determinants of practice and knowledge. To include patients in medical history was one obvious way to establish a sub-discipline separate from the 'histories written by (old) doctors' – not to exclude the technical aspects of medicine but to mesh them with the more obviously 'social'.[2]

Rich resources were then to hand, both in universities and in reformist or radical politics. From medical sociology and anthropology, and from the remains of the 'social medicine' movement among clinicians,[3] aspirant social historians of medicine learned that professional medicine was but a small part of the experience and management of illness. Medical sciences and professional practices were not the major reasons for the falls in death rates from the later nineteenth century: wider histories of society and the economy helped explain the difference. Medical encounters involved more than patients in 'sick roles', for patients had their own ways of understanding illness, which social scientists around medical schools were busily uncovering. The growth of 'social history' as a subfield of history[4] also strengthened the conviction that medical history must include the politics of medical practices and institutions, and American examples, from Richard Shryock's textbook to regional studies, were useful here.[5] With the People's History movement – a late extension of the British workers' education movement – there came a particular interest in working-class experiences and constructions, including experiences of patients.[6] From feminist history, came a focus not just on maternity and sexuality, but on domestic management of illness.[7] And from oral history and anthropology were borrowed the techniques for engagement with these themes, especially for the history of the still living.[8] So resourced and armed, patients' history could aspire to be part of a History of the People, including lost voices, resistances and alternative histories. Links between these various academic and political tendencies were built through voracious reading and attendance at national meetings, but also through local connections between various university departments and related groups.[9] The History Workshop movement was particularly important for engagement with local historical communities, achieved through its annual conferences.

As noted, colleagues in all these tendencies wanted to go beyond 'old docs' history' and the 'medical model'. They wanted to include active patients and to write histories of interest to patients; but how then were patients to be envisaged? The next sections of the chapter examine more closely some of the extensions of the medical model then available from the social fringe of clinical medicine, to see how doctors might see their patients. I also look at the suppositions of activists then emergent around the NHS, especially around issues of women's health. I show why, at this time, the constructions of ordinary patients by reformist historians did not draw on models of powerful patients; though available, these were locked in the past. In the second half of the paper I show how this historical model of 'patients as patrons' became relevant and potentially useful, how that utility may be enlarged, and how patient power may fit with other desiderata and powers in the force-fields of medical politics.

## Psychologising and socialising the patient

The usual 'medical model' was made most explicit by its critics, and by those doctors who for various reasons wanted to go beyond it. It was not necessarily the limit of the post-war medical vision, but it was the normal starting point for discussion of medical encounters. Disease was basically lesions of tissues, or perhaps functional disturbances, revealed by clinical signs such as heart murmurs, to which the clinician rather than the patient had privileged access. That kind of medical model was seen to have developed through the nineteenth century, with substantial reinforcement from antiseptic and aseptic surgery, germ theory, synthetic pharmaceuticals, and interwar clinical physiology. Antibiotics from World

War II had confirmed its power, and post-war governments in Britain and the US had invested heavily in lab-based medical research and in hospital medicine.

A fuller view of medicine in society might have said more about 'public health' and social investigations, but the NHS was hospital centred. Germs and deep poverty seemed mostly to be yesterday's problems; the main focus now was on the improvement of clinical services. One way to improvement lay through new constructions of patients. In 1984, David Armstrong, a general practitioner turned medical sociologist, looked back to three ways in which new kinds of patients had come into being, two of which involved psychiatry.[10]

Before the NHS, psychiatry had largely comprised a tiny market in private practice alongside a large national industry of asylum super-vision. From the 1960s, the use of asylums began to decline and com-munity care became the dominant paradigm, and in some regions, psychiatric departments were included in general hospitals. More generally, 'community doctors' were meant to know more than they once had about mental diseases.[11]

Psychiatry for non-psychiatrists was also part of the revival of general practice in the NHS. Post-WWII GPs had often been seen as second rate doctors, fit only to filter patients towards hospitals, and to receive back such cases as still required medical care. In 1948, the social medicine movement had offered clinicians a variety of legitimations for medical work 'in the community', but for various reasons that movement failed, including its attempts to reform general practice.[12] Greater success came, however, with the promotion of psychiatric sensibilities as part of the role of a GP or family doctor. Michael Balint's writings on this theme were crucial for many of the elite who set up the Royal College of General Practitioners. These new GPs, and the new psychiatrists, analysed the psycho-dynamics of illness and medical care, and saw some diseases as rooted in the social and psychological circumstances of the patient.[13] The circumstances in question now *were* the diseased systems, and the words of the patient were the major evidence of these disorders. The taking of a patient history became idealised as a com-municative interaction, in which the doctor's role might also be analysed.

These new psychosocial models in the later twentieth century might be seen as biographical extensions of medical models, constructed to meet the needs of 'new' GPs who defined themselves by their under-standings of patients rather than diseases.[14] But psychiatrically attuned medics were not the only people interested in patients' views and their felt needs, nor in doctor-patient communication. From the 1950s, sociologists and a few anthropologists studied such matters by surveys and interviews in the USA and (perhaps especially) in Britain, and their work provided a third psycho-social construction of patienthood.

In the new world of the NHS, and perhaps also in the more public parts of the US systems, the efficiency and proper operation of health services seemed to depend on the behaviour of patients. Patients' attitudes might then be studied in the hope of making patients more efficient – visiting surgeries when they ought, staying away when they could safely do so, and obeying instructions about preventive and therapeutic measures. The problematic addressed by much medical sociology was how to ensure that the 'sick role' was adopted when, and only when, it was clinically appropriate. But that enquiry broadened into studies of patients' beliefs and communication.[15]

We can note here that questions of patient attitude were especially important for cancer. Elizabeth Toon's recent work has explored the dynamics of the research and education programmes led from the 1930s by Ralston Paterson, the influential Director of the Christie cancer hospital in Manchester.[16] The initial aim of these programmes was to find ways of 'reducing ignorance' in the hope of shortening the delay between patients noticing key symptoms and their resort to a doctor. But it transpired that many patients knew only too well the significance of their symptoms: from their informal knowledge of other such cases, they had concluded that little could be done; they avoided doctors and hospitals to delay the inevitable disruption of their lives and families.

As this example suggests, the post-war British public was often envisaged as ignorant of science. Scientific professionals employed what was later called the 'deficit model' of public understanding – if patients were to make better use of services the public mind needed to be filled with reliable information.[17] But many medical sociologists and anthropologists tended to construe medical interactions rather more symmetrically than had the sponsors of their research. Patients' worlds and language were not just extensions of the 'medical model'; patients' narratives offered an alternative way of construing medicine. Patients were of interest in their own right, and we were all (potential) patients.

### Critique, patient activism and the NHS

The introduction of the NHS, as well as increasing total access, had widened the social range of 'public' medical services. From the National Insurance Act of 1911, public medicine had been for 'workers' (or children or paupers); now it was for the population at large. (The private sector was then very small, and partly accommodated within NHS hospitals; it expanded in the later 1970s when a Labour government pushed it out of NHS premises). This widening class-base for public medicine would seem to have underlain some of the critiques of medicine, academic and lay, which emerged in the 1960s. Their authors were of a generation raised to assume that public services were for everyone, including themselves, and that the public should have its say. Medicine was not to be left to doctors; the public interest needed to be *represented*.

The NHS, for the first time in British history, gave general access to medicine, free at the point of need. It was a relatively classless service which still carried the strong communal spirit with which some voluntary hospitals had been associated.[18] But it did not carry forward those few pre-war initiatives which had given real power to workers' associations, nor did it give much voice to health workers other than doctors; the voice of medicine was almost always that of the hospital consultant.[19] The 1970s NHS, in the eyes of many critics and analysts, was a universal public service which still acted as if its patients were the dependent poor. That professions and public authorities were paternalist was surely not new, but the articulation of objections by other kinds of professionals was more of a novelty. Such critiques were usually intended to make space for more open or democratic mechanisms, perhaps guided by professional visions beyond those of clinicians.

By the first major reorganisation of the NHS, in 1974, as Chris Ham pointed out, the 'public interest' was becoming identified with the patient interest, with users of the service.[20] The Patients Association, founded in the early 1960s, was in part a reaction to revelations about clinical research without patient consent, but it soon came to be complemented by 'patients' groupings for particular disease categories. Some of these were new associations organised by lay sufferers or their relatives; some had been started by interested doctors and then 'democratised'. Consumerism did not define the 'people', but by the 1970s it had become integral to notions of the 'public' in debates about public services.[21]

These developments can be placed alongside the emergence of consumerism more generally. The Consumers' Association was founded by Michael Young, a Labour politician and sociologist best known for his studies of the working class; but the Consumers' Association was essentially middle class, more a part of the new 'issue' politics than of class politics. Its links with social historians' growing interest in consumption were chiefly through studies of the eighteenth century, where most of the actors were of the middling classes and above,[22] not so much with the more radical politics of medical oppression and resistance which found its historical roots in nineteenth-century politics, including gender politics.[23]

## Women, children and hospitals

In some ways the most effective pressure group for the reform of particular services was the women's movement, which went beyond critique to the provision of alternative services, some of which reached across class. Images from feminist history played a part in these constructions which sought to 'demedicalise' reproduction and womanhood more generally. These movements are well known for the 1970s but it is worth noting that one of the first challenges to traditional hospital disciplines came in the 1950s, for children's services.

Interestingly, the campaign for the well-being of children in hospitals, and especially for increasing the roles of mothers, also had roots in psychiatry. This development was stimulated by research, and especially a film, by James Robertson, who worked at the Tavistock Clinic, and drew on the wartime work of Anna Freud and the attachment theories of John Bowlby. Robertson's revelations of distress among hospitalised infants were taken as insulting by many professionals. Nevertheless, these revelations prompted an enquiry, published in 1959 and headed by Sir Harry Platt, the Manchester orthopaedic surgeon, long familiar with the local women-led babies hospital. In 1961, Robertson initiated a campaign around his films, and his cause was soon taken up by a group of mothers who met in a park in Battersea, London. Guided by Robertson, they established Mother Care for Children in Hospital, which later become the National Association for the Welfare of Children in Hospitals (NAWCH). Post-1974, the movement was able to use the 'consumer representation' bodies of the restructured NHS to gain attention for its message, and NAWCH campaigned successfully not just for relaxation of visiting rules and involvement of families, but for play provision and educational facilities in hospital as well as better transport provisions for parents.[24]

Seen in long perspective, this psychologisation of the child patient might be seen as an extension of medicalisation, like the Balint analysis of the primary care patient. But the concomitant involvement of mothers, and then of families, makes clear a different politics. About the normal behaviour of the child, and the fulfilment of its needs, the mother could claim more knowledge and skills than attributed to doctors; and in important respects the environment of the hospital came to be judged in relation to, and in continuity with, the home. The 'isolation' and disciplining of the child patient, which had been seen as part of the cure, thus became a systematic defect of hospitals, but one which good parenting (and ancillary professions such as play leaders) could help minimise. The active involvement of mothers of sick children seems to precede the feminist critique of hospital midwifery, and the well-women campaigns. The connection merits further study, as does the connection of feminist medicine with the disability movement; so too do the cumulative effects of these movements on hospitals more generally, including expectations of the doctor-patient relationships.[25]

But how were these evidences of patient activism connected with current medical history? What did historians have to say about patient power, and especially about doctor-patient relations?

### People not patrons

In considering the origins of 'the medical model' and the passive patient, most historians of medicine were at one with most medical sociologists. They pointed to the hospitals of post-revolutionary Paris, and referred to Ackerknecht's study and the better known work of Foucault to stress the subordination of patients to technical routines. In Britain from the 1970s, it was part of the critique of the medical model, as discussed above. It could be a spur to extensions of the medical model, or to its rejection, but the presentation of medicine in Foucault's *Birth of the Clinic* offered no obvious positive model as an alternative. The work was primarily about the conditions of emergence of biomedicine and surgery, and in the tradition of Bachelard and Canguilhem, these were understood as developments 'sanctioned' by history, part of the history of truth rather than ideology. Much medical history and sociology was then associated with Foucauldian approaches; its subjects were power and control, not patient perspectives or choice.

The historiography of women patients was mostly of medicalisation; likewise for the deaf and the disabled who were then becoming active. But the reformist claims here were for wider views of normality, rather than for active patienthood. Explorations of patient power tended to focus on the politicised medical sects of the nineteenth and early twentieth century, including their various continuities with heterodox medicine from the 1960s.[26] These involved vernacular models of health and illness, and the autonomy of the patient, to be sure – but mostly in refusing regular medicine rather than amending it.

There was, however, an alternative historical model of the patient. It involved 'bedside' medicine, which was partly the invention of Ackerknecht, drawing on accounts of elite medicine in Ancient Greece and

Rome, mostly written in Baltimore, USA, by German speaking émigrés and refugees, notably Edelstein, Temkin and Sigerist.[27] In Britain, this view of medicine was best known from two papers by Nicholas Jewson, a young sociologist at Leicester University, who linked bedside medicine to aristocratic and gentry patronage, exploring this relationship for eighteenth-century England.[28]

Leicester was then notable among British sociology departments for the presence of émigré German scholars keenly interested in history – including Norbert Elias. It would be useful to learn more about the local roots of Jewson's work, but its explicit derivation was from the Parisian Marxism of Louis Althusser which related modes of knowledge to the conditions of their production. The British Althusserian school also produced other interesting studies of biomedical sciences (e.g. Paul Hirst on Claude Bernard), but for various reasons they lacked direct progeny.[29] But though Jewson himself took up the sociology of sport (another interest of Elias), his two medical papers, continued to provide a useful organising principle for social history of medicine courses, especially in and about Britain.

The Jewson scheme has usually been presented as a model of transition – from the bedside medicine of the eighteenth century to the hospital medicine of the early nineteenth, to which laboratory medicine was later added. In bedside medicine, disease was a disturbance of the patient's life, discussed in a language of symptoms, in terms shared by patients and their doctors. There was little separation of mind and body, or ontological independence of different diseases. Disorders were characterised by the intensity of symptoms, their shifting bodily locations, their changes over time, and the response to changing circumstances, including regimen and therapeutics. Crucially, patients differed from each other not just in their circumstances and their constitutions but in their philosophies, preferences and goals. Doctors needed to understand this variety, and in as much as doctors were dependent on their more powerful patients, they needed to find an accommodation with the preferences of the patrons through discussion. That summary fitted easily with Roy Porter's many sympathetic explorations of eighteenth-century medicine, his strong interest in the history of consumption (in both senses), and his aversion to overly Foucauldian perspectives. For Porter, the pluralism of enlightenment medicine was more an epistemic and social virtue than a preventative of science.[30] Yet historians very rarely pursued this model into the nineteenth and twentieth centuries, to ask about patients' expertise and the forms of their patronage in the centuries of organised medicine and organised resistance.[31]

That this restricted 'historical' view of bedside medicine was then little challenged might, in retrospect, have reflected the increasing divisions between medical history and clinical medicine. Britain then lacked both academic philosophy of medicine and the strong tradition in 'history of ideas', which in the US linked professional history of medicine more easily with philosophical questions about the clinic and with classical traditions.[32] In British history of medicine there was then comparatively little about ethics or professional responsibilities – these were still left to doctors, rather than ethicists, and the few doctors involved in social history of medicine were mostly interested in public health.[33] The models of professionalisation in social history of medicine came mostly from sociological histories of modernisation and industrialisation; patients were commonly identified with the working classes and seen as individually powerless.[34] As noted, the then emergent history of consuming was differently structured, but historical research on medical markets tended to focus on the early modern period; it seemed marginal to the ongoing medicine of a country which had rejected medical markets in favour of a tax-funded health service led by salaried doctors.

Perhaps this simplistic historical vision of successive kinds of medicine in successive periods was also shaped by a sociological reductionism which followed from determined contextualisation. Historians tended to focus on particular periods and places to show how the dominant forms of medicine were related to dominant aspects of context. If different sites and styles were compared, it was usually for contrasts rather than continuities: thus patronage in the eighteenth century and 'industrialised medicine' in the nineteenth. In the historiographic world of Kuhnian revolutions, any kind of cumulation was liable to seem Whiggish, and attempts to talk historically to the present ran the risk of presentism.

Yet had the question been raised, who could deny that the patient power of 'bedside medicine' continued to be important long after 1800 – whenever and wherever patients exercised choice and their patronage mattered to their doctors. From the American frontier in mid-nineteenth century to Harley Street in Edwardian London, one could find medicine adapted to patients' perceived wishes, whatever the level of technical sophistication involved. Indeed, John Harley Warner showed how American doctors returning from France adapted their Paris medicine to American markets, and Chris Lawrence how the medical grandees of 1920s London subordinated medical science to a medical mode focused on the individuality of patients (and doctors).[35] But such historiography, though much appreciated, seemed then to show how some doctors resisted novelties and preserved older traditions; it seemed a social history of residues rather than of medical improvement.

For the most part, historians saw patients as subordinated; their imagined remedies usually lay in public representation to make services more responsive. But at the end of David Armstrong's paper of 1984, he did raise questions about the 'conditions of possibility' under which a patient-centred understanding might become a force in practice (not merely an addition to the medical model), and he noted questions which were then emergent – about patient representation but also about patients' rights and community politics. Today there is much more to be said about these issues – partly because the writing of history has changed, but especially because of radical changes in the presumed roles of NHS patients.

## Part Two: 'Making patients central' in the neoliberal NHS and in historiography

The key shifts in patient power over the last 25 years in England were in part a result of leftist critiques and popular movements to which we have already referred. But they were also the results of a resurgence of neoliberalism from the late 1970s under Conservative governments (and then New Labour), which undermined both professional and communitarian models. In their stead, business models were promoted, both by direct private sector involvement in health services and by remodelling government services according to the routines of business – stressing accounting, new public management, goal setting etc. Professional power, once concentrated in doctors, has been dispersed across many health service professions, and doctors have lost autonomy (though not salaries), in part because seniority in medicine was not strongly associated with management roles in the way which has become familiar for nursing. Much power has moved to lay managers, now trained in schools of management rather than post-war style departments of public administration. Political decisions, e.g. about allocation of resources between competing demands are now routinely framed by management consultancies, and policy formation seems to be elided with the enlarged management roles of central government. The distinction between public and private providers has been gradually erased by the insistence that performance is the only politically important feature of supply mechanisms. Thus the identity of the NHS as a public service and as a contribution to social justice has come to rest more on its funding from general taxation, than the public forms of its delivery.[36]

The justifications of this new, unstable NHS have relied heavily on the identification of public interest with that of individual consumers. There was, for example, little discussion of the historically important community-cohesion functions of hospitals (even when there was a Minister for Communities), and little about the conditions necessary to foster professional autonomy and development. Central management of the system was focused on deliverable outcomes, and more often on the reduction of grievance than on any overall measures of outcome. Hence the huge importance of reducing waiting lists, even when it compromised some other outcomes.

From around 2000, reforms emphasised patient choice. Patients seeing a GP and needing secondary care were offered a choice of hospital facilities, including private sector providers (with payment at fixed tariffs from the NHS). Some critics have seen the emphasis on choice as a diversion from questions of overall standards, and many see it as deliberately favouring private sector provision, but no one can doubt the centrality of 'choice' in the ideologies of successive governments since 2000.

How then are patients expected to make decisions between and within services? The model, 'expert' patient is now to be informed by the media and the Internet, and perhaps through patients' associations. Indeed, compared even to the 1980s, huge quantities of information are accessible to certain kinds of patients. That the 'problem patient' of 2000 is 'over-informed' may be a stereotype, but many patients, not least for cancer, and perhaps especially breast cancer, are indeed very well-informed; they have read narratives and web sites of other patients, and researched the benefits and side-effects of various possible treatments.[37]

### The empowered patient and the uses of history

This new focus on patients and the explosion of material for and by patients may transform the conditions of the historical study of contemporary medicine.[38] It greatly magnifies two characteristic differences between recent and more remote foci for historical study: a) that more information tends to be available for the more recent, and b) that this information is shared by informants and potential audiences. In our cancer project, we have drawn heavily on interviews; and concurrent work on orthopaedic patients used electronic resources extensively.[39] As historians dealing with common medical conditions at a time when the media are full of patients' tales, we more easily identify with patients and envisage our results as potentially useful to them. More generally, these shifts in the meanings of 'public' since the 70s pose new questions about medical history, its audiences and its potential utility.

For English policy makers and some patients, of course, one potential benefit of history follows rather directly from the rapid and repeated service reorganisations. These have so drained organisational memory from the NHS that collective amnesia is severe.[40] Hardly anyone in the service now knows how the NHS functioned in 1995, let alone 1975 or 1955. So part of the practical historical task is to restore narrative and allow assessment of both the motors and impacts of organisational and functional changes.

But at another level and time scale, how might patients use the 'narratives' which now feature prominently in 'medical humanities', often dealing with past times but not focused on the dynamics of change? The study of literature is promoted within medical school curricula as likely to increase the sensitivity of future doctors to patients' views. Such studies, whether from literature or history, may also be an important resource for patients themselves, linking directly with much of the self-help literature. One task for historians may be the comparative analysis of such material across times (and places), so that we can all learn more about synchronic and diachronic differences in patients' experiences and in common understandings of illness and medicine.

Yet in as much as patients, and perhaps especially cancer patients, are encouraged to think of their treatments and services as evolving rapidly, so too their understandings may be explicitly historical in the dynamic sense. When the durations of treatments are comparable with the durations of trials and service modifications, a patient may 'live' the technical changes in a very direct way; and when cancers appear to run across generations, the evolution of services and treatments may become part of family history as well as medical history.[41] Historians, here, can add analyses and critiques to the 'time-lines' about treatments which are anyway in circulation, and they can integrate patients' accounts and wider contexts. For new cancer treatments, as for NHS changes, these histories may be eventful even when short.

Our utility as historians, however, may also derive from our 'long-views', and whether we can connect presents and pasts in ways which run deeper than common discussions. Do we, for example, have ways of envisaging patient histories over long periods, which can also serve for the present – both as means of analysis and as a mode of communication? Can we offer historical views of medicine which include both the 'medical model' and the empowered patient; and which might provide some purchase for the constructive critique of both these models?

## Reasserting patients in medical history and humanities

From the 1980s, the more that NHS patients came to be seen as consumers, the more plausible and necessary seemed histories in which 'sick-persons' did not disappear – and where their continued presence was more than a lingering, or an occasional resurrection required by psychological medicine or service designers. This historiographical opportunity was becoming clear by the mid-1980s as the proliferation of cultural studies emphasised the complexities of medicine, the many voices, and the variations across synchronic contexts. But while each strand of what is now called medical humanities, whether from literature, anthropology or science studies, has since brought offerings to the medico-biographical feast, these contributions have been connected more by the dialects of theory than by historical sociology. Indeed, the present British stress on medical humanities arises in part from this profusion. We seem to collect approaches in the hope they may hybridise, and thus help humanise medicine.[42]

Medical sociology no longer links medical history with studies of the present, in the ways outlined above for the years around 1980. Though some of the 'social sciences in relation to medicine' have continued to thrive, they seem more 'siloed', with quantitative studies in demography, epidemiology, equality studies and service evaluations, and with qualitative studies of cultures. In the eyes of present-day organised medicine, the most important of the 'social' sciences is health economics, which has largely replaced 'social administration and policy' as the referent discipline for policy makers. As clinical research has come to be dominated by clinical trials, so health economists have been called on to help assess benefits and costs. But few people in medical history or medical humanities understand such matters. Bioethics too has grown enormously, but again, much of it seems too technical and normative for the tastes of cultural historians; they engage with it, if at all, through historicising critiques of the discipline and its professionalisation.[43]

The academic configurations around medical history are now very different from those described for circa 1980. Generally they are more about cultural differences and less about causes (in either sense).[44] But perhaps this diversification of approaches and disciplines should suggest that we look harder for general models which will help the continued development of medical socio-historical disciplines in ways one might call 'vertebrate' – rather than, shall we say 'coralline'. Here I borrow 'vertebrate' from Harold Perkin's formulation of social history as being centrally about major changes in social (and cultural) structures. As this present article suggests, it is not too hard to see, in outline, the recent interactions of politics, economics, medicine and history. We may differ as to interpretations, but as publicly-funded historians we have a responsibility to address major issues and contemporary questions, including those about our own work.

### Ways of knowing and patient histories

In this respect, history of medicine has seemed to me a privileged discipline, compared to most history of science and most social history. It had the opportunity and incentives to continue to incorporate the proliferation and shifts of cultural perspectives, including patients' views, as it also developed deeper historical understandings of technical practices. One such move was to relativise the Jewson scheme, and to think of biographical medicine (rather than just bedside medicine) as a persistent form of medical knowledge and politics, interacting and perhaps including 'hospital medicine' and 'laboratory medicine'. By thus imagining a series of *contested cumulations* in medicine, we might escape both from over-simple models of succession and from the proliferation of uncoordinated case studies.

This model of cumulative ways of knowing was developed for medicine in the later 1980s, then extended across the history of science and technology and first published in 1993.[45] Since then it has been variously developed,[46] but in this essay we can return again to the home ground as it were, to the ways in which 'bedside medicine' in the eighteenth century might be seen as both culture and nature, and how one might extend the biographical model to include nineteenth- and twentieth-century developments of analytical and experimental medicine.[47] The next sections briefly develop this view for cancer studies, discussing the content of biographical understandings and their relations with technical understandings, before returning to issues of expert patients and patronage.

### Biographical medicine, including analysis and experimentation

Like most forms of biography, and indeed most forms of common knowledge, 'bedside medicine' made no clear separation between the cultural and the natural, say between symptoms and their cultural resonances. As historical analysts, however, we can benefit from that distinction, partly to underline the importance of the cultural and affective, but also to give status to 'natural history' in medicine, as pursued both by clinicians and less formally by patients. Much improvement, in medicine as in other technical fields, stems from careful observations, classifications and 'craft work', rather than esoteric analysis. If we recognise that 'natural history' (and associated craft-work) continues to be important in medicine, we help make room for expert patients and the sharing of information about experiences. Thus we also usefully undermine the exclusive equation of 'medical science' with the less accessible analyses and experimentation associated with hospital and laboratory medicine. Patients can create natural-historical knowledge as well as collect it, and though they may not have access to hospital tests, they can learn how to use them, or refuse them, and to situate them in their own biographies.[48]

Earlier in this essay I roughly equated the 'medical model' with 'hospital medicine', Paris style. More strictly, the hospitals of early nineteenth-century Paris saw the addition of analytical methods to the existing corpus of medical natural-history (along with old and new cultural perspectives). Pathological anatomy was the foundation of the wider medical model which later included many types of analytical medicine (and then those I would call 'constructively experimental'). From gross pathology and blood chemistry to antibodies and genetic testing, analytical methods have proliferated and cumulated; you can now find them all in hospital laboratories. They have become a central aspect of most biographical medicine.

These analytical methods often allowed rationalisations of treatment, by providing classifications for the collection of data, from the 'numerical methods' of Pierre Louis in Paris to the data-banks of present day Evidence Based Medicine.[49] For example, as I have argued elsewhere, radiotherapy, especially from the 1930s, may be seen as a paradigm of analytical and rationalised medicine: in the centralised treatment centres, a high proportion of the cancers treated were 'staged', given standard doses, and entered into large data banks which served to assess the efficacy of the treatments.[50]

Nowadays there are many complex series of analytical tests for cancer; and therapies may further define the cancer, by 'working' or not. Such matters are understood by many patients as well as doctors. Patients may now move through protocol sequences which produce both diagnoses and outcomes. The patients may be in statistical classes which are defined analytically, and they may be in corresponding social groups, and each class or group may have their modes of life and their meanings. But the intersections of these classes may come to produce individualised trajectories, with probabilities at each turn, as it were; perhaps paradoxically, the multiplication of 'reductive analyses' may thus help constitute new forms of individuality. But however that aspect developed, the experience of analysis, in all these forms, must surely be one focal point of our shared understandings of modern cancer. Not just, how do patients come to terms with a diagnosis and prognosis, but how do they live the time-courses of tests and therapy, reactions, risks and potential recurrence?

Recent cancer medicine, however, goes far beyond analysis. Modern clinical experimentalism, born in the 1940s out of laboratory experimental method and 'causal' statistics', has from the 1980s become paradigmatic for research-linked treatment.[51] Experimentalism here is a formalisation which incorporates and goes beyond the analytical in stressing novel interventions. To be sure, the hope placed in novelties by desperate patients and doctors is one of medicine's oldest stories – but in recent experimentalism it is statistical, and the patient (like the clinician) is 'blinded'. The hopes and fears which attach to best-practice routines and to novelty are thus reconfigured, and so is the patient's potential contribution to the public good. To say that this experimentation incorporates analysis and natural history, and cumulates their meanings (for patients and for doctors), seems to me a useful formulation for historical analysts – and for patients and doctors.

But if we shift our focus from medical models to the patient's perspective, we can see all these cumulated aspects of medicine subordinated to the cultural, in a wider understanding of 'biographical medicine'. Life as a patient may be partly a matter of protocols or trials, but it is also about hair loss or constipation, about complex sequences of hospital visits and domiciliary care; about endless conversations about symptoms, feelings, expectations, fears, and the complexities of getting through the week.[52]

As suggested earlier, any extension of medical models to include more biography might sometimes serve to 'medicalise' everyday life. But with different power relations, the mix can work the other way, for example enfolding anatomical or physiological analysis in wider biographical frames where the patient or carer may share the expertise and make the choices. Like Jewson's eighteenth-century aristocrats, they may exchange understandings with physicians and gain accommodations to their preferences. Perhaps thinking of them as patrons may thus be more useful than the usual stress on choice and consumerism.

## Consumers and/or patrons

The advocacy of 'patient choice' in neoliberal regimes might well be seen as promoting a more biographical model than had been common in the post-war decades, but in eliding medical choices with consumerism, the market view underplays the more political aspects of the relationship. The biographical model, by contrast, usefully also includes the negotiating power of the patient as patron. Choice as such may not be central; perhaps any doctor could provide the answers asked for, provided the patient was in a position to keep on asking.

Though patient patronage may now seem an antique notion, contrasting with more 'modern' social (and market) relations, its relatives would seem to be in fashion again, at least among (some) political scientists. For example, in his *Between Politics and Science*, David H. Guston gives an extended constructivist and historical analysis of US science policy, using 'principal-agent theory' to explore the changing relations between scientists and government, including the calls for accountability and quality regulation which parallel those in medicine.[53] The first three of his paired examples of principal/agent are patron/performer, customer/green-grocer, and patient/doctor. In such pairs, the 'principal' is meant to have the power, but the 'agent' has more knowledge and therefore discretion. Market relations, it would seem, are here modelled as a subdivision of patronage by stressing the technical knowledge of the seller in the context of the power of the principal.

I know too little of political theory, but this model seems relevant here. We tend to think of the new model-patient as a consumer, but this wider reading of patient as principal or patron seems to fit the long history of elite-patient medicine and the shorter history of commoners supposed to be empowered by the NHS. It gives us a useful handle on longer medical history, and may encourage us to think of how patients may use doctors, as well as how they choose them. It can thus revise the post-Second World War literature on patient-doctor interactions which I discussed in connection with the early NHS. In this extended medical model, communication was about patients getting the message; in more symmetrical accounts of communication, and especially in patron-based accounts, it would be about how best a doctor can understand and meet the patient's demands.

Like all ideal types, the patient-as-patron model is an analytical tool to be used alongside other types and other tools. Patients manifestly differ in their ability to be successful patrons, and some empowerment surely requires collective action by patrons, as in patient associations or local (more or less) democratic agencies. These aspects are clear in the chapter by Keating and Cambrosio, which highlights the collective power of patients in negotiating clinical trials requiring their collaboration. The recurrent problems of medical politics may be shifted – but they are certainly not solved. How these demands may be balanced in practice has been illustrated here in the paper by Zuiderant-Jerak, Bal and Berg on the redesign of Dutch cancer services.

Though 'patient as patron' has to be seen as a moral ideal alongside effectiveness, equity, economy and respect for professional judgement, it may nonetheless add a new dimension to the politics of public medicine and to its analysis, including medical history. It provides another normative standard for medical relations – as human relations, not simply as biological 'outcomes'.[54] Systems of medicine are to be judged, in part, by the quality of the information and services they make available to patients, and the ways in which they allow informed choices to be delivered. And, as Valier's paper suggests, information here must mean more than advertisement: doctors and public authorities have a responsibility to provide evidence across a range of options, including effects on public costs as well as private benefits. Indeed, one of the achievements of the recent NHS has been the development of balanced assessment mechanisms and the publicising of their judgements.[55] Narrowly read, these may be seen as technocratic controls, rationing medical services in the face of consumer demand, but here again a wider and more political perspective may transcend market models. In as much as public patients may exercise their patronage in the public interest as well as the private, technical assessments may be a means of building consensus about 'reasonable' limits for public expenditure or the utilisation of professional skills. The structures and dominant ideologies of the present do not preclude attention to the issues of solidarity which motivated critiques of the NHS and of medical historiography in the 1970s.

## In conclusion

We might then, for some purposes, wish to effect reversals in the historical medical model which ran from patient patronage to hospital medicine and its extensions. Though we certainly need to understand the growth and differentiation of 'medical models' and the subordination of patients which often accompanied them, these are not the whole story, even for public or state medicine. If they sometimes seem so, it is because we often look from the doctors' side – and because the biographies of most patients are difficult to recover. But where the patients are individually or collectively powerful, in the past or present, we can sometimes give substance to their standpoint, including the experience of medical institutions. We can construct richer accounts of the lives which they lived, with the more technical histories in parenthesis as it were.

Whether under these conditions, the technical 'inclusions' are simply 'black-boxed' by the patient, or probed and reconfigured, will depend on the circumstances, and ideally on the patient's choice. Some patients may choose to rely heavily on their doctors and limit their own enquiries. So too, some historians may choose to chart the histories of technical medicine, treating patients as *products* of health care rather than consumers. Which does not mean that these historians, or the doctors, or the patients, are careless of the patients' interests – all may take a pride in outcomes which are good for the patients, even where patients had little or no input. The broader challenge, however, and the one of most potential use to patients for most purposes, will include the patient's view as primary to that of the technical agents.

If the history of patients is a crucial part of medical history, history *for* patients may be a way of seeing all the medicine which we study and re-present in and to our own times. It may also be a way of pulling together our technical, ethical and cultural histories of medicine, broadly understood, into a capacious but coherent frame.

## Acknowledgements

I gratefully acknowledge discussions with, and comments from, David Armstrong, Margaret Pelling and my Manchester colleagues, including Flurin Condrau, Stephen Harrison, Carsten Timmermann, Elizabeth Toon, Stephanie Snow, Duncan Wilson and Michael Worboys.

## Notes

1  Porter and Porter 1989.
2  For analogous developments in the US, see Reverby and Rosner in Huisman and Warner (ed.) 2004.
3  Porter 1997.
4  For a pioneer's story see Perkin 2003.
5  For examples, Shryock 1936 and Bonner 1991.
6  The meetings can be traced in *History Workshop Journal*.
7  For example O'Hara 1977, Davin 1978, Figlio 1978, Versluysen 1980, Lewis 1980, Eccles 1982, Oakley 1984.
8  A standard guide was Thompson 1978.
9  As an example, I can note the academic constellations around Manchester University in which we delighted circa 1980. The main grouping was in the reformist department of Public Health/Social Medicine/Community Medicine (name changes were common), which encompassed social medicine, medical sociology, feminist approaches, and some medical anthropology. Michael Bury was researching the self-construction of arthritis patients; the anthropologist

Jean Comaroff worked with a psychiatrist on the meanings of childhood leukaemia as it became (more) curable; Joyce Leeson and Judith Gray, both public health doctors, wrote on feminist approaches to medicine. History of medicine was then newly added to the History of Science and Technology department at UMIST, whilst critical science studies, including an interest in 'science for the people', was centred in the department of Liberal Studies in Science, where Edward Yoxen was already exploring the politics of the new genetics. Welfare history was to be found in History, along with the anthropological history of the Africanists led by Terry Ranger; and welfare history was central to the Department of Social Administration – reflecting the historical orientation of that still hopeful discipline which in London, Manchester and Britain more generally, had been designed to underpin the post-war welfare services. (For examples of their work in the late 1970s, see Bury 1982; Comaroff and Maguire 1981; Leeson and Gray 1978; Yoxen 1983; Ranger and Slack 1992; George and Wilding 1994). Local history and oral history thrived at Manchester Polytechnic, especially around Bill William's Manchester studies programme.

10    Armstrong 1984.

11    For a view by a Salford psychiatrist and historian, see Freeman 2005. Hugh Freeman was one of the pioneers of psychiatry in District General Hospitals, on which he wrote extensively. In their development of 'community' facilities, some of these 'district psychiatrists' represented a major extension of the 'medical model' – the hegemony of which collapsed with the growth of generic social work, community mental health nursing and generic management of community services.

12    E.g. Perry 2000.

13    Balint 1957.

14    For the lack of psychiatry in interwar general practice Britain, see Tudor Hart 2000.

15    Helman 1984.

16    Toon 2007.

17    See Pickstone 2000, Chapter 8.

18    For voluntary hospitals in Lancashire towns, and for a 1970s 'political ecology' approach to the history of hospitals, see Pickstone 1985.

19    See for example, Webster 1998. For a sympathetic account of workers' medicine by a scholar who later headed a right wing think-tank see Green 1985.

20    Ham 1977.

21    For a very useful account of consumers, see Mold 2010. For a very useful review of patients' history, emphasising its difficulty and the variety of patients' experiences, see Condrau 2007.

22    For example, Brewer and Porter 1993.

23    For example, Cooter 1988.

24    Peg Belson (1993) outlines the story. Also see the chapter by Barnes in this volume, and historical material on the websites of the charity Action for Sick Children (the later name for NAWCH) and especially their CDs: 'An Interview with Sir Harry Platt' (1986), and '50 years after Platt: Where are we now?' (2009). The case is well analysed in a recent seminar paper by Mold 2011.

25    For the History Workshop meeting in Salford in 1989 we organised a session on learning disability, thinking we were usefully extending the scope of the

medical history strand; we were surprised to be asked why the session was not part of minorities history.

26 E.g. Cooter 1988.

27 For a selection, with a historical introduction, see Temkin 1977. Note that Temkin's *Janus*-faced doctors continued to form a significant section of the American Association of the History of Medicine; this became less true of the (British) Society of the Social History of Medicine.

28 Jewson 1974 and 1976.

29 Hirst 1975.

30 See Cook 2007.

31 But see the important later paper by Jacyna 1992.

32 That the Jewson model was less popular among American social historians than British may be due to the emphasis on aristocracy which appeared to put it outside American medical history.

33 Wilson 2011b.

34 For an example, also from Leicester, see Waddington 1984.

35 Warner 1985; Lawrence 1985.

36 On the recent history of the NHS, see Klein 2010 and Harrison and McDonald 2008.

37 See Mold 2010.

38 On materials created for patients by patients, see also the chapter by Barnes in this volume.

39 See Anderson *et al.* 2007.

40 Pollitt 2000.

41 See the chapter by Baines in this volume.

42 For one history of the history of medicine in the UK see Pickstone 1999.

43 Rothman 1991; Cooter 1995; Wilson 2011a and 2011b.

44 As historians of medicine in Manchester now, we have fraternal relations with medical ethics, equality studies, health service management, cultural history and innovation studies, but little remains of the public health department and there is no obvious focal point for social studies in medicine. The most inter-disciplinary of our historical ventures is a group around the recent history of the NHS, which connects historians with clinicians and policy academics; and a parallel group around the recent history of mental health. We have also sought interactions with scientists and clinicians (see for example, Jones and Pickstone 2008, and Jones and Snow 2010). For technical histories we have collaborators among the social scientists of innovation and among historians of science and medicine, often outside Britain. Those in Britain are mostly less interested in 'contemporary history', partly because so many are now in history departments, with weaker links to biomedical audiences.

45 Pickstone 1993a, 1994 and 2000.

46 Pickstone 2009, 2011a, 2011b and 2011c.

47 Pickstone 1993b.

48 For excellent collections of patients' histories, see the website of the Oxford-based charity DIPEx at http://www.healthtalkonline.org (as of September 2011).

49 Rosser Matthews 1995, Daly 2005.

50 Pickstone 2007b.

51 See the chapter by Keating and Cambrosio in this volume, and Keating and Cambrosio 2007.

52  See the chapters by Baines and Timmermann in this volume.
53  Guston 1996.
54  For doctor as friend, see Lain Entralgo 1969, for normativity in history of STM see Pickstone and Worboys 2011.
55  For NICE see Harrison and McDonald 2008, esp. Chapter 7.

## Works Cited

Anderson J., Neary F. and Pickstone J. V. 2007, *Surgeons, Manufacturers and Patients. A Transatlantic History of Total Hip Replacement*, Basingstoke: Palgrave.

Armstrong D. 1984, 'The Patient's View', *Social Science and Medicine*, 18, 731–44.

Balint M. 1957, *The Doctor, His Patient and the Illness*, London: Churchill Livingstone.

Belson P. 1993, 'Children in Hospital', *Children and* Society, 7, 196–210.

Bonner T. N. 1991, *Medicine in Chicago, 1850–1950: A Chapter in the Social and Scientific Development of a City*, 2nd edition, Chicago: University of Illinois Press.

Brewer J. and Porter R. (eds) 1993, *Consumption and the World of Goods*, London and New York: Routledge.

Bury M. 1982, 'Chronic Illness as Biographical Disruption', *Sociology of Health and Illness*, 4, 167–82.

Comaroff J. and Maguire P. 1981, 'Ambiguity and the Search for Meaning. Childhood Leukaemia in the Modern Clinical Context', *Social Science and Medicine, Part B. Medical Anthropology*, 15, 115–23.

Condrau F. 2007, 'The Patient's View Meets the Clinical Gaze', *Social History of Medicine*, 20, 225–40.

Cook H. J. 2007, 'Roy Porter and the Persons of History', in R. Bivins and J. V. Pickstone (eds) *Medicine, Madness and Social History: Essays in Honour of Roy Porter*, Basingstoke: Palgrave, 14–24.

Cooter R. (ed.) 1988, *Studies in the History of Alternative Medicine*, London: Macmillan.

Cooter R. 1995, 'The Resistible Rise of Medical Ethics', *Social History of Medicine*, 8, 257–70.

Daly J. 2005, *Evidence-Based Medicine and the Search for a Science of Clinical Care*, Berkeley: University of California Press.

Davin A. 1978, 'Imperialism and Motherhood', *History Workshop Journal*, 5, 9–66.

Eccles A. 1982, *Obstetrics and Gynaecology in Tudor and Stuart England*, Kent, Ohio: Kent State University Press.

Figlio K. 1978, 'Chlorosis and Chronic Disease in 19th-Century Britain: The Social Constitution of Somatic Illness in a Capitalist Society', *Social History*, 3, 167–97.

Freeman H. 2005, 'Psychiatry and the State in Britain', in M. Gijswijt-Hofstra, H. Oosterhuis and J. Vijselaar (eds) *Psychiatric Cultures Compared: Psychiatry and Mental Health Care in the Twentieth Century: Comparisons and Approaches*, Amsterdam: Amsterdam University Press, 116–40.

George V. and Wilding P. 1994, *Ideology and Welfare*, Brighton: Harvester.

Green D. G. 1985, *Working Class Patients and the Medical Establishment: Self-help in Britain from the Mid Nineteenth Century to 1948*, New York: St Martins Press.

Guston D. 1996, 'Principle-Agent Theory and the Structure of Science Policy', *Science and Public Policy*, 23, 229–40.

Ham C. J. 1977, 'Power, Patients and Pluralism', in K. Barnard and K. Lee (eds) *Conflicts in the National Health Service*, London: Croom Helm, 99–120.

Harrison S. and McDonald R. 2008, *The Politics of Health Care in Britain*, London: Sage.

Helman C. 1984, *Culture, Health and Illness*, London: John Wright and Sons.

Hirst P. 1975, *Durkheim, Bernard and Epistemology*, London: Routledge.

Jacyna S. 1992 'Mr Scott's Case: A View of London Medicine in 1825', in R. Porter (ed.) *The Popularization of Medicine, 1650–1850*, London: Psychology Press, 252–86.

Jewson N. 1974, 'Medical Knowledge and the Patronage System in Eighteenth-century England', *Sociology*, 8, 369–85.

Jewson N. 1976, 'The Disappearance of the Sick Man from Medical Cosmology', *Sociology*, 10, 225–44.

Jones E. L. and Pickstone J. V. 2008, *The Quest for Public Health in Manchester: The Industrial City, the NHS and the Recent History*, Manchester: Manchester NHS Primary Care Trust, in association with CHSTM, University of Manchester.

Jones E. L. and Snow S. 2010, *Against the Odds: Black and Minority Ethnic Clinicians and Manchester, 1948 to 2009*, Manchester: Manchester NHS Primary Care Trust, in association with CHSTM, University of Manchester; Distributed by Carnegie Publishing, Lancaster.

Keating P. and Cambrosio A. 2007, 'Cancer Clinical Trials: The Emergence and Development of a New Style of Practice', *Bulletin of the History of Medicine*, 81, 197–223.

Klein R. 2010, *The New Politics of the NHS*, 6th edition, Oxford: Radcliffe.

Lain Entralgo P. 1969, *Doctor and Patient*, New York: McGraw Hill.

Lawrence C. J. 1985, 'Incommunicable Knowledge: Science, Technology and the Clinical Art in Britain, 1850–1914', *Journal of Contemporary History*, 20, 503–20.

Leeson J. and Gray J. 1978, *Women and Medicine*, London: Tavistock.

Lewis J. 1980, *The Politics of Motherhood: Child and Maternal Welfare in England, 1900–1939*, London: Croom Helm.

Mold A. 2010, 'Patient Groups and the Construction of the Patient-Consumer in Britain: An Historical Overview', *Journal of Social Policy*, 39, 505–21.

Mold A. 2011, 'Re-Positioning the Patient: Patient Groups, Consumerism and Bioethics in Britain, 1960s–1970s', unpublished paper given at CHSTM, the University of Manchester, 21 November.

O'Hara M. J. 1977, *Elizabethan Dietary of Health*, Lawrence, Kansas: Coronado Press.

Oakley, A. 1984, *The Captured Womb: A History of the Medical Care of Pregnant Women*, Oxford: Basil Blackwell.

Perkin H. 2003, *The Making of a Social Historian*, London: Athena Press.

Perry M. 2000, 'Academic General Practice in Manchester Under the Early NHS: A Failed Experiment in Social Medicine', *Journal of the Social History of Medicine*, 13, 111–29.

Pickstone J. V. 1985, *Medicine in Industrial Society. A History of Hospital Development in Manchester and its Region, 1752–1948*, Manchester: Manchester University Press.

Pickstone J. V. 1993a, 'Ways of Knowing: Towards a Historical Sociology of Science, Technology and Medicine', *British Journal for the History of Science*, 26, 433–58.

Pickstone J. V. 1993b, 'The Biographical and the Analytical: Towards a Historical Model of Science and Practice in Modern Medicine', in I. Löwy (ed.) *Medicine and Change: Historical and Sociological Studies in Medical Innovation*, Paris: Les Editions INSERM, John Libbey, 23–46.

Pickstone J. V. 1994, 'Museological Science? The Place of the Analytical/Comparative in Nineteenth Century Science, Technology and Medicine', *History of Science*, 32, 111–38.

Pickstone J. V. 1999, 'The Development and Present State of History of Medicine in Britain', *Dynamis*, 19, 457–86.

Pickstone J. V. 2000, *Ways of Knowing. A New History of Science, Technology and Medicine*, Manchester: Manchester University Press.

Pickstone J. V. 2007a, 'Working Knowledges Before and After c. 1800: Practices and Disciplines in the History of Science, Technology and Medicine', *Isis*, 98, 489–516.

Pickstone J. V. 2007b, 'Contested Cumulations: Configurations of Cancer Treatments through the Twentieth Century', *Bulletin of the History of Medicine*, 81, 164–96.

Pickstone J. V. 2009, 'From History of Medicine to a General History of "Working Knowledges"', *International Journal of Epidemiology*, 38, 646–9.

Pickstone J. V. 2011a, 'Sketching Together the Modern Histories of Science, Technology and Medicine', *Isis*, 102, 123–33.

Pickstone J. V. 2011b, 'A Brief Introduction to Ways of Knowing and Ways of Working', *History of Science*, 49, 235–45.

Pickstone J. V. 2011c, 'Afterwords', *History of Science*, 49, 350–74.

Pickstone J. V. and Worboys M. 2011, 'Introduction', *Isis*, 102, 97–101.

Pollitt C. 2000, 'Institutional Amnesia: A Paradox of the "Information Age"?', *Prometheus*, 18, 5–16.

Porter D. and Porter R. 1989, *Patient's Progress: Doctors and Doctoring in Eighteenth Century England*, Cambridge: Polity Press.

Porter D. (ed.) 1997, *Social Medicine and Sociology in the Twentieth Century*, Amsterdam: Rodopi.

Ranger T. and Slack P. 1992, *Epidemics and Ideas: Essays on the Historical Perception of Pestilence*, Cambridge: Cambridge University Press.

Reverby S. M. and Rosner D. 2004, '"Beyond the Great Doctors" Revisited: A Generation of the "New" Social History of Medicine', in F. Huisman and J. H. Warner (eds) *Locating Medical History. The Stories and their Meanings*, Baltimore and London: The Johns Hopkins University Press, 167–93.

Rosser Matthews J. 1995, *Quantification and the Quest for Medical Certainty*, Princeton: Princeton University Press.

Rothman D. J. 1991, *Strangers at the Bedside: A History of How Law and Bioethics Transformed Medical Decision Making*, New York: Basic Books.

Shryock R. H. 1936, *The Development of Modern Medicine. An Interpretation of the Social and Scientific Factors Involved*, Philadelphia: University of Pennsylvania Press.

Temkin O. 1977, *The Double Face of Janus and Other Essays in the History of Medicine*, Baltimore: Johns Hopkins University Press.

Thompson P. 1978, *The Voice of the Past: Oral History*, Oxford: Oxford University Press.

Toon E. 2007, '"Cancer as the General Population Knows It": Knowledge, Fear, and Lay Education in 1950s Britain', *Bulletin of the History of Medicine*, 81, 116–38.

Tudor Hart J. 2000, 'Going to the Doctor', in R. Cooter and J. Pickstone (eds) *Medicine in the Twentieth Century*, Amsterdam: Harwood Academic, 543–58.

Versluysen M. C. 1980, 'Old Wives' Tales? Women Healers in English History', in C. Davies (ed.) *Rewriting Nursing History*, London: Croom Helm, 175–99.

Waddington I. 1984, *The Medical Profession in the Industrial Revolution*, London: Gill and Macmillan.

Warner J. H. 1985, 'The Selective Transport of Medical Knowledge: Antebellum American Physicians and Parisian Medical Therapeutics', *Bulletin of the History of Medicine*, 59, 213–31.

Webster C. 1998, *The National Health Service. A Political History*, Oxford: Oxford University Press.

Wilson D. 2011a, 'Creating the "Ethics Industry": Mary Warnock, In Vitro Fertilization and the History of Bioethics in Britain', *BioSocieties*, 6, 121–41.

Wilson D. 2011b, 'Who Guards the Guardians? Ian Kennedy, Bioethics, and the "Ideology of Accountability in British Medicine"', *Social History of Medicine*, 24, advance access: doi:10.1093/shm/hkr090.

Yoxen E. 1983, *The Gene Business. Who Should Control Biotechnology*, London: Pan Books.

# Index

CPSIA information can be obtained
at www.ICGtesting.com
Printed in the USA
LVOW04*1254240116

472003LV00007B/419/P

9 781137 272072